STUDY GUIDE

to accompany

Wallace

BIOLOGY

The World of Life

Sixth Edition

Prepared by
Steven J. Muzos
Austin Community College

HarperCollins*CollegePublishers*

To Anne, Amanda, Nathan, Briley, and Katie

Study Guide to accompany Wallace's BIOLOGY: THE WORLD OF LIFE, Sixth Edition

ISBN: 0-673-46626-4

92 93 94 95 96 9 8 7 6 5 4 3 2 1

TABLE OF CONTENTS

PART FOUR REPRODUCTION AND DEVELOPMENT 91

PART FIVE SYSTEMS AND THEIR CONTROL 114

PART SIX BEHAVIOR AND THE ENVIRONMENT 165

TO THE STUDENT

Welcome to the World of Biology! By reading this section of the study guide, you have shown that you are interested in living things and are ready to take an active part in learning the material for your biology course. Congratulations, you are off to an excellent start!

IMPROVING YOUR STUDY SKILLS

In order to have a successful learning experience in your biology course, you must have an effective plan for learning before, during, and after class. Consider using **all** of the following activities:

talk to your instructor, either during class or outside of class ... he or she will see that you are interested, and you will both learn from the exchange

attend class regularly and **take part** in all class activities ... the more active you are the more likely you are to learn

prepare for class ahead of time ... you will have a chance to **listen and think** about what is going on, instead of merely writing furiously

take notes as you need them ... and revise them regularly

become part of a **study group** ... research has shown that if you **talk about the material** you are likely to understand it and remember it

use this study guide ... it has many suggestions, summaries, and practice exercises for every chapter

make use of any **tutoring facilities** on your campus whenever you want additional help

plan to spend time **studying this course outside of class** several days each week ... write a specific schedule if that helps you ... but don't wait to cram before exams

use your text ... scan the major headings, read it, write comments in the margins, study the art work, answer the questions, learn the summaries, and use brief underlining

complete your assignments on time and turn them in ... use your instructor's comments to help yourself learn

find ways to **connect and apply** what you are learning **to what you already know** ... make an effort to tie the course material to the other aspects of your life make and use **flash cards** for the key terms and main concepts of each chapter ... details are described below.

Educational research has shown that the items listed above can be the **components of a successful study strategy for you**. Plan your approach to the course wisely, and have fun!

IMPROVING YOUR THINKING SKILLS

Biology students are expected do much more than just learn and repeat facts. As you are studying biology, you will be required to use some of the same thinking skills that scientists employ when they develop and use biological information. In other words, you will be thinking like a biologist.

In order to do this, it is important to develop your higher order thinking skills. These include problem solving, decision making, critical thinking, and creative thinking. Although you may not always be aware of it, you use these in your daily life. This study guide provides you with many opportunities to practice and improve these important skills with biological examples.

You will be asked to consider the difficult question, "Why?" In attempting to answer this, it will be important for you to be able to sort through examples of poor reasoning, biases, and unnecessary information. You will also have to formulate and support explanations with facts and reasoning that are appropriate to the question that you are considering. Occasionally it will be necessary for you to choose between what may seem to be very similar alternatives, or to offer your own alternative solutions. In addition, you will have to assemble your information, organize it, and present it in a clear and concise style.

Even with all of this effort, it should become obvious that neither the professional biologists nor you know all of the answers, and you will have to tolerate uncertainty at times. That can be frustrating, but it is what takes place in the study of science. Hopefully, you will begin to ask "Why?" on your own, and use your skills to seek answers.

USING THIS STUDY GUIDE

Following are brief summaries of the features that you will find in each chapter of the study guide. Also included are suggestions on how to make the best use of these features.

CONNECTIONS

This section is a preview of the chapter. It ties together material that you have already learned with information that is coming up in the current chapter. With this, you will have a way to connect the new ideas to what you already know.

OBJECTIVES FOR THIS CHAPTER

These are specific statements that describe what you should know and be able to do when you have finished with the material in the chapter. The objectives in this study guide are similar to, yet more specific than, those in the text. They will help you to anticipate what to look for as you work through the material.

CHAPTER OUTLINE

The outline presents the main topic items in a organized form. This will help you see how the items are related to one another.

CONCEPTS IN BRIEF

These are concise descriptions and summaries of the ideas that are covered in each of the *sections* (as are shown in the chapter outline) of the text.

KEY TERMS

Key terms are noted at the end of each chapter in your text, and they are repeated here. However, the lists of terms in this study guide include a notation of the page where the term is introduced in the text, and may also include additional terms of importance.

You should make **flash cards** for each of the Key Terms listed for the chapter. Carry the cards around with you and quiz yourself and the members of your study group on a regular basis. Use cards (such as 3" x 5") and write the term on one side.

ARTIFICIAL SELECTION

1 Front of flash card

On the other side, write a definition in your own words, *not* just what you can copy from the glossary in the text. Along with your definition, add an example and a current application of each term. If you like, you can also add a note regarding the location of the term in the text.

DEF:	The breeding of living things, by humans, in order to produce offspring with certain desired characteristics
EX:	Pets, livestock, crops, flowers
APPL:	Many types of fancy goldfish are bred for their looks, even though their characteristics would not help the fish to survive in the wild

2 Back of flash card

EXERCISES

Completing this section will give you the chance to test your knowledge and thinking skills after finishing each chapter. Answers for all of the exercises are found in the ANSWERS TO CHAPTER EXERCISES section (see p. 207). Before you look at the answers, however, be sure that you have written your answers in the study guide itself. Don't cheat yourself of the chance to learn from your own right and wrong written answers. Making and correcting your own mistakes in the study guide will help you clarify your ideas.

SUMMARY

This section reminds you to look at the chapter summary in the text, and helps you connect that material to the overall themes of the chapter.

HINTS

If you are having trouble with a particular exercise, look in this section. You will find reminders that will help you avoid common errors on some of the EXERCISES.

IN CLOSING

It is important that you take an active role in your own learning for this course. Following the suggested study strategies that are presented above will help you to have a more positive and successful experience this term. I wish you the best of luck as you learn about the fascinating World of Biology.

Part One Life and the Flow of Energy

CHAPTER 1

A BRIEF HISTORY AND THE ENCHANTED ISLES

CONNECTIONS

You have probably noticed that there are many different kinds of living things in the world today. In addition, there are numerous species that have become extinct. Like scientists throughout the ages, you may have wondered how, or even why, such variety developed.

This chapter examines the development of modern scientific explanations for the diversity of life on the earth. Pay attention to the influence that various scientists had on one another's thinking as you see ideas develop and change over time.

OBJECTIVES FOR THIS CHAPTER

When you have finished Chapter 1, you should be able to:

1. Describe Charles Darwin's personal and educational background.
2. Give examples of Darwin's activities as a naturalist during the voyage of the Beagle.
3. Summarize the progress and direction of scientific discovery prior to the 1700s.
4. Describe the scientific contributions of Eratosthenes, Copernicus, Galileo, and Newton.
5. Compare the views that scientists held on the changeability of species from before Darwin's writings with those after Darwin's writings.

6. Describe and explain Jean Baptiste de Lamarck's views of how evolution occurred.
7. Outline the ideas of Charles Lyell and explain how his ideas influenced Darwin.
8. Outline the ideas of Rev. Thomas Malthus and explain how his ideas influenced Darwin.
9. Explain the process of natural selection as described by Charles Darwin.
10. Explain how Darwin's thoughts on natural selection were molded by his observations as a naturalist.
11. Compare the descriptions of how living things change as described by Lamarck and Darwin.

CHAPTER 1 OUTLINE

I. The Voyage of the Beagle
II. The History of an Idea
III. The Beginnings of Biology
IV. Time and the Intellectual Milieu
V. The Development of Darwin's Idea of Evolution
   A. The Impact of Lyell
   B. Puzzling Change and Variation
      1. The Galapagos Islands
   C. The Impact of Malthus
VI. Natural Selection

CONCEPTS IN BRIEF

The Voyage of the Beagle

When Charles Darwin set sail on the H.M.S. *Beagle* in 1831, he was well educated but had not been an impressive student. His father was disappointed in the educational progress that Charles had made, and was not enthused about the voyage that his son was beginning. Charles, on the other hand, was very excited about following his passion to be a naturalist.

Darwin had the opportunity to spend a great deal of time ashore in South America. He was a careful observer and collected a wide variety of scientific specimens.

The History of an Idea

During the centuries before 200 A.D., scientists (who were also philosophers) devoted great effort to the study of the earth and its life forms. Many important discoveries were made during this time. Then, for approximately 1,000 years, western science became intermixed with and dominated by religion. Instead of being able to test ideas, people were required to follow doctrine. Much of what was known by scientists was suppressed, and science seemed to make little progress.

In spite of persecution, many scientists continued to work and write. From 1200 - 1700 A.D., people such as Copernicus, Galileo, and Newton made important contributions to the knowledge of the day. More and more people began to think that the world was understandable and predictable. Because of these feelings, the fear of the unknown and the power of magic over people's lives began to decrease.

The Beginnings of Biology

Scientists in the 1700s began to question the ideas of special creation. They suggested that not all of the species on the earth had been created in their current forms. Both the French naturalist Buffon, and Erasmus Darwin, stated that some organisms had changed their form through time.

In the early 1800s, the French naturalist, Lamarck, examined fossil evidence and said that some older species had even given rise to newer species. Even though that was correct, we now know that his explanations of *how* these changes occurred were not accurate. There is no evidence to support Lamarck's ideas of organisms generating structures by need, and improving structures by use. Despite his errors, Lamarck made an important contribution to the scientific debate concerning the changeability of species.

Time and the Intellectual Milieu

England was more conservative than France in the early and middle 1800s. Many scientists and nonscientists still felt that species were fixed in form.

The Development of Darwin's Idea of Evolution

Many people influenced the development of Darwin's ideas of evolution. Charles Lyell, a geologist, said that the world was very old, and that the geological processes of the past were still operating on the earth. From this, and from his own study of geology and fossils, Darwin realized that living things would have had long periods of time to change their forms. These realizations on Darwin's part made him very uncomfortable, since he started his voyage as a firm believer in the permanence of species.

Throughout his journey, Darwin became aware of the natural variation of organisms in the environment. He also noticed the changes that could be seen in species as he traveled from one location to another in South America. When he reached the Galapagos Islands he was struck by the fact that many types of animals were *not* found on *all* of the islands. During his stay on the Galapagos Islands, he made extensive collections of small birds known as finches. Years later he would realize that the finches had originally come from South America, and that different environments on the islands had influenced their changes in form.

When Darwin returned to England in 1836, he was welcomed as an important scientist, but he had a great deal of work yet to do. He read an old essay by Malthus, and realized that natural forces (such as limited resources) must be keeping the size of natural populations in control. Otherwise the world would be at least knee deep in everything! He reasoned that one important factor controlling populations would be the ability of an organism to survive long enough to reproduce.

Natural Selection

The development of Darwin's idea of natural selection was influenced by his experience with artificial selection, such as the breeding of livestock to promote certain characteristics. No one in England knew much about genetics, but it was widely known that the traits of the parents were passed to their offspring.

After reading the essay by Malthus, Darwin realized the important role that competition and limited food supplies played in natural populations. He added these ideas to his growing concept of natural selection. He felt that the environment set the conditions under which everything must live. Finally, he said that the organisms that were the most successful at surviving *and* reproducing would pass along their traits to more offspring than would the organisms that were not as successful in their reproduction.

By not knowing how variations appeared in living things (as explained by modern genetics), his explanations were not complete. In spite of that, it was clear to many people that he had put forth a very important biological concept. Alfred Russel Wallace wrote to Darwin with essentially the same idea before Darwin had a chance to publish his work. In the end, Darwin received credit for the concept of natural selection because he had been collecting evidence and writing notes and letters about it for several years prior to that time.

For a variety of reasons, many scientists, clergy, and lay people were very uncomfortable with the ideas of natural selection as set forth in Darwin's *The Origin of Species*, published in 1859. Darwin's health was poor by this time, and he relied heavily on his many strong defenders and a great deal of scientific evidence to support his work. There was no better explanation available at that time that could cause his ideas to be rejected by science.

KEY TERMS

Be sure to make and use flash cards for all of these terms. Suggestions are found in "To The Student" at the beginning of the Study Guide. Page numbers refer to the text.

artificial selection	25	special creation	13
changeability of species	11	variation	26
natural selection	26		

EXERCISES

Check your understanding of Chapter 1 by completing the following exercises. Answers begin on p. 207.

1. Arrange the following events in sequence, from earliest to most recent.

   a. Lyell publishes the first volume of *Principles of Geology*.

   b. Wallace sends his ideas of natural selection to Darwin.

   c. Darwin begins his voyage on the *Beagle*.

   d. Darwin starts to study theology at Cambridge.

   e. Lamarck proposes the development of animal structures through their use by the animal.

   f. Darwin reads the essay by Malthus.

   ______ ______ ______ ______ ______ ______
   Earliest   Most Recent

True or False Questions

Mark each question either T (True) or F (False).

____2. Both Lamarck and Darwin said that species change through time.

____3. Galileo's ideas were popular and widely accepted in his time.

____4. Darwin was generally considered to be an excellent student.

____5. Malthus wrote that it was important to control human populations.

____6. Darwin's concept of natural selection made it clear that the strongest animals are usually the ones that survive and reproduce.

Matching Questions

Each question may have more than one answer; write all of the answers that apply. Answers may be used more than once.

____7. felt that all species were created in their present form.

____8. said that species may change through time.

____9. began a system to classify living things.

____10. felt that competition was important in the formation of species.

____11. noticed that older fossils seemed to be simpler than newer ones.

____12. said that acquired traits were inherited.

____13. wrote about the influence of food supplies on populations.

a. Erasmus Darwin
b. Buffon
c. Malthus
d. Lamarck
e. Linnaeus

Multiple-Choice Questions

Circle the choice that is the best answer for each question.

14. As Darwin traveled, he noticed that
    a. there was not very much evidence of modern geological activity.
    b. variation was common among groups of organisms.
    c. it was clear that special creation had happened in South America.
    d. the species on the Galapagos were the same as those found on the mainland of Ecuador.
    e. all of the above.

15. Charles Lyell stated that
    a. the earth was much older than had been thought.
    b. ancient geological processes were no longer operating.
    c. the process of evolution did not apply to the physical characteristics of Earth.
    d. geological evidence proved that animals did not change through time.

16. A time period in which there was very little scientific progress was
    a. 200 A.D–1200 A.D.
    b. 1300–1700.
    c. 1700s.
    d. 1800s.

17. The finches that Darwin collected were finally classified according to
    a. the length of their tails.
    b. the island on which they lived.
    c. their color.
    d. the shape of their beaks.
    e. all of the above.

18. Darwin eventually decided that finches were originally
    a. blown to the islands by the wind.
    b. brought to the islands by sailors.
    c. the result of special creation.
    d. carried to the islands by the tortoises.
    e. a good example to support Lamarck's theories.

19. As part of his concept of natural selection, Darwin said that variations appear
    a. randomly.
    b. in response to what the organism needs.
    c. because of genetic mutations.
    d. whenever there are earthquakes.

20. When Darwin published *The Origin of Species*
    a. it sold out in one day.
    b. he meant it to be his final work.
    c. the scientific community agreed with his ideas.
    d. Alfred Russel Wallace helped him write it.
    e. all of the above.

Fill in the Blanks

Complete each statement by writing the correct word or words in every blank.

21. Darwin stated that if an organism had a competitive edge it may have a (an) _______________ chance to survive.

22. Lamarck said that animals could change their characteristics due to _______________, but Darwin's evidence showed that variation occurred _______________. Both men thought that these changes could be _______________.

23. Darwin was impressed by the unique animals of the _______________ Islands, off the coast of_______________. He made collections of _______________, which are small birds.

24. In order to pass on their traits, animals must _______________ long enough to _______________.

Questions for Further Thought

Write your answer for each question in the space provided.

25. How are the processes of natural selection and artificial selection *different* from each other?

26. Darwin's family objected to him going on the voyage aboard the *Beagle*. What were some of their objections?

27. a. Why was Darwin reluctant to change his mind about species being fixed in form?

    b. What kinds of evidence finally convinced him to change his mind?

SUMMARY

Study the seven items listed in the SUMMARY on p. 29 of the text. You should have noticed that many different scientists contributed to the idea that species may not be fixed in one form for all time. Darwin's concept of natural selection acted as a unifying theme in biology. It explained that natural forces in the environment act on and affect living things. Understanding this makes it clear that a possible result is that groups of living things may show a change in their characteristics over time.

HINTS

Read these suggestions before you check your answers to the EXERCISES.

For question 2: This only asks about the end results, not how the results were achieved.

For question 4: The question asks about Darwin as a *student*, not as a researcher.

For question 6: The key word here is "strongest."

For question 17: This asks about what was eventually (finally) used, not just what data were first collected.

For question 19: The concepts of genetics were not well understood in Darwin's time.

CONGRATULATIONS . . . YOU HAVE COMPLETED CHAPTER 1 !!!

CHAPTER 2

SCIENTISTS AND THEIR SCIENCE

CONNECTIONS

In spite of what television may imply, scientists do not have the answers to all of the questions that people ask. Chapter One provided you with examples of how scientists influence one another. You are about to see some of the ways that individual scientists go about doing the day to day work of solving scientific problems. As you read, look for the human side of science and the roles that scientists play in today's world.

You may never have had any trouble deciding whether or not a particular thing was alive. Biologists, however, have had a difficult time defining life. Use this chapter to begin refining your personal criteria for distinguishing living things from nonliving things. Be aware that your criteria may change throughout the course.

OBJECTIVES FOR THIS CHAPTER

When you have finished Chapter 2, you should be able to:

1. Recognize and write examples of deductive and inductive reasoning.
2. Explain how deductive and inductive reasoning are employed to formulate a hypothesis.
3. Describe the scientific method and describe examples of its use.
4. Differentiate between hypothesis and theory.
5. Explain how a controlled experiment is used to test a hypothesis.
6. List six features that characterize living things.
7. Describe how the concepts of teleology and phenomenology affect the observations of scientists.
8. Discuss some special social concerns that confront today's scientists.

CHAPTER 2 OUTLINE

I. Inductive and Deductive Reasoning
   A. Science Trends and Shifting Logic
   B. Hypothesis, Theory, and the Scientific Method

II. The Scientist as Skeptic
III. Defining Life
    A. The Trap of Teleology
    B. Science and Phenomenology
IV. What Biologists Do For a Living
V. Science and Social Responsibility

CONCEPTS IN BRIEF

Inductive and Deductive Reasoning

In the past, there have been arguments over which type of reasoning was the best to use for good scientific work. Today, both inductive reasoning and deductive reasoning are important to modern scientists, and are used together in the process of establishing facts. There is also no single acceptable "scientific method" used by all scientists. Instead, investigators use a combination of procedures to arrive at answers to questions.

Observations of natural events add to our existing knowledge and generate additional questions to solve. A hypothesis, or possible explanation for the observations, will be used to make a prediction about what might happen in a *different* situation. The predictions must be testable. Scientists design experiments that allow them to test the predictions of their hypothesis. The results of the experiments help scientists learn if their predictions are accurate.

Experiments often have a control. This involves setting up what may look like a second experiment, but it is really part of the original work. The control part of the experiment is set up to be exactly like the first part of the experiment, however the one thing that is being tested will be left out. This lets the experimenter see the effect of that one thing. Designing the experiment (or series of experiments) with a proper control that allows you to determine an accurate conclusion is an important skill.

Theory is defined differently in science than it is by non-scientists. Scientific theories are established ideas that are supported by much experimental evidence.

The Scientist as Skeptic

Scientists are human, like all of us, but they tend to share the characteristic of being skeptical. When they are doing good scientific work, they require evidence before they will accept ideas. It is important that scientists examine each other's work very carefully and repeat it in order to arrive at the most accurate answers.

Defining Life

A precise definition of life is simply not available. Instead, biologists usually list various characteristics that living things have in common. Then they agree that anything which shows all (or sometimes most) of those features is alive. The text lists six such qualities as its criteria for life.

We do not know whether or not other kinds of living things can make conscious plans or goals. Therefore it is important to avoid using teleology when we describe their activities. Living things do whatever they do, without regard to our explanations, so it is easy to find exceptions to the rule. However, when scientists interpret what organisms do, it is important that personal biases are avoided. Otherwise the result might be inaccurate interpretations because of phenomenology.

What Biologists Do For a Living

Biologists all study living things. The text (page 44) shows examples of the variety of ways this may be done. Even the traditional divisions between different sciences are no longer as clear as they once were.

Science and Social Responsibility

This section presents many questions and few answers. It is clear that scientists must be active and concerned members of the world community. It is not clear what (if any) special role they must play because of the fact that they are scientists with access to new or specialized information. Each scientist must make his or her own professional and personal decisions on a wide variety of issues. It is important for *you* to begin to clarify your own views on this topic as well, including a definition of your own role in the world of today and the world of the future.

KEY TERMS

Be sure to make and use flash cards for all of these terms. Suggestions are found in "To The Student" at the beginning of the Study Guide. Page numbers refer to the text.

biologist	43	deductive reasoning	33
control	35	experiment	35
hypothesis	35	teleology	42
inductive reasoning	33	testable predictions	35
phenomenology	42	theory	35, 37
scientific method	35		

EXERCISES

Check your understanding of Chapter 2 by completing the following exercises. Answers begin on p. 207.

1. Arrange the following activities in sequence, from what is usually first to what is usually last. Some stages in this process may not appear as possible choices.

   a. Scientists accept a new theory.

   b. A conclusion is made from experimental data.

   c. Observations are made.

   d. A prediction is made.

   e. A hypothesis is developed.

   f. An experiment is designed.

_______ _______ _______ _______ _______ _______

First Last

True or False Questions

Mark each question either T (True) or F (False).

____2. Scientists should allow their personal biases to influence their interpretations of data.

____3. Isaac Newton said that scientists should rely on inductive reasoning.

____4. A good hypothesis should lead to testable predictions.

____5. A scientific theory is a tentative explanation that still needs to be tested.

____6. It is important for scientists to question the methods and conclusions of other scientists.

Matching Questions

Each question may have more than one answer; write all of the answers that apply. Answers may be used more than once.

____7. Implies that a goal is behind the actions of an organism.
____8. When specific statements are derived from general observations.
____9. Results in personal bias influencing an interpretation of data.
____10. Scientists probably rely most heavily on this for developing scientific principles.
____11. "Some birds nest high in trees to have a safer place to raise their young".

a. inductive reasoning
b. deductive reasoning
c. phenomenology
d. teleology

Multiple-Choice Questions

Circle the choice that is the best answer for each question.

12. Exploring new areas on Earth
    a. is still possible today.
    b. is not a job for biologists.
    c. has not been possible since the 1920s.
    d. is only possible in the tropics.
    e. both a and d are correct.

13. Scientists have
    a. better personal opinions than nonscientists.
    b. unique social responsibilities.
    c. clearly defined social responsibilities.
    d. all of the above.

14. A scientific theory
    a. is really an educated guess by a scientist.
    b. is similar to a scientific personal opinion.
    c. is a common result of teleology by scientists.
    d. is based on much scientific research.
    e. all of the above.

Fill in the Blanks

Complete each statement by writing the correct word or words in every blank.

15. If you gather data and use those data to generate a general statement, you are using __________ reasoning. You could also make a (an) __________, from which you could make predictions. These predictions must be __________.

16. Living things are made of small units called __________.

17. In your daily life, you probably look for __________ in order to tell if something is alive.

18. Organisms must be able to __________ and __________ over time to changes in their environment. In a more immediate sense, it is possible to observe the fact that many living things __________ or react to the world at any moment.

Questions For Further Thought

Write your answer for each question in the space provided.

19. Describe the role of a control in an experiment.

20. Pick three characteristics of life and explain why each one is important to living organisms.

    a.

    b.

    c.

SUMMARY

Study the seven items listed in the SUMMARY on p. 50 of the text. You should have noticed that there is not one fixed method of problem solving that is used by all scientists. Rather, scientists rely on a combination of observations, experience, reasoning processes, experimentation, and communication with other scientists to develop and change their hypotheses and theories. In addition, you may have noticed that when scientists arrive at answers, they always end up with even more questions! It is important to accept the fact that uncertainty is common in science. It is what scientists deal with all of the time.

HINTS

Read these suggestions before you check your answers to the EXERCISES.

For question 2: This concerns bias, not professional opinion.

For question 5: It is asking about a theory, not a hypothesis.

For question 10: This says mostly, not always.

For question 14: Remember to use the *scientific* definition of theory.

For question 18: Read question seventeen first.

For question 19: Do not *define* a control here. Discuss its function in the process of science.

For question 20: This is not asking for definitions. Consider how each characteristic may be beneficial to the organism.

CONGRATULATIONS . . . YOU HAVE COMPLETED CHAPTER 2 !!!

CHAPTER 3

THE CHEMISTRY OF LIFE

CONNECTIONS

By now it is probably too late for you to drop this course, so if you took biology in order to avoid chemistry you're out of luck! In the last chapter you were correctly warned that a biologist must be part chemist and part physicist. Chapter Three introduces you to the idea that chemical activity is at the heart of living systems (which is true, and not just a bad pun) and their daily activities. Watch for the applications that are used as examples throughout the chapter. Do not get so caught up in the fine details that you forget how it all applies to living things. You will find that the chemistry of organisms will reappear throughout the text.

This chapter introduces a great number of new terms to learn and use. The sciences have a large specialized vocabulary that may seem clumsy to use at first. If you make and use your flash cards you will have an easier time applying the terms correctly. This will help you in the rest of the course, since many of these basic terms and concepts will reappear in later topics. Hopefully you will see that using the correct terminology makes it easier to discuss the subject.

OBJECTIVES FOR THIS CHAPTER

When you have finished Chapter 3, you should be able to:

1. Distinguish between an atom and an element.
2. List the six elements (with their symbols) that comprise 99 percent of living matter.
3. List the subatomic particles.
4. Relate the structure of an atom to its chemical properties.
5. Describe how electrons are arranged around a nucleus.
6. Describe how the organization of electrons in nuclear orbitals affects an element's reactivity.
7. Name three types of chemical bonds and state how each is formed.
8. Explain how elements combine in chemical reactions.
9. Distinguish between endergonic and exergonic reactions.
10. Describe the special bonding properties of carbon.
11. Explain the mechanisms by which enzymes act as catalysts in chemical reactions.

12. List the four major groups of biological molecules and describe the components, structure, and properties of each group.

CHAPTER 3 OUTLINE

I. Atoms and Elements, Molecules and Compounds
II. CHNOPS
    A. The Structure of an Atom
III. How Atoms May Vary
    A. Isotopes
    B. Ions
IV. Electrons and Their Behavior
    A. Oxidation and Reduction
    B. Electrons and the Properties of the Atom
V. Chemical Bonding
    A. Ionic Bonds
    B. Covalent Bonds
    C. Hydrogen Bonds
VI. Chemical Reactions
    A. Energy of Activation
    B. Enzymes
    C. Energy Changes
VII. The Molecules of Life
    A. The Magic of Carbon
    B. The Role of Water in Forming Large Biological Molecules
    C. Carbohydrates
    D. Lipids
    E. Proteins
        1. Amino Acids
        2. Peptides
        3. Protein Structure
    F. Nucleic Acids

CONCEPTS IN BRIEF

Atoms and Elements, Molecules and Compounds

All living and nonliving things are made of extremely small units called atoms. These atoms join together to form larger units called molecules. What you can actually see around you are large quantities of these. If a material is made of just *one* kind of atom it is called an element, each of which has a unique symbol to represent it. If a material contains two or more kinds of atoms it is a compound.

CHNOPS

There are six elements that make most of the "stuff" of living things. The atoms of these (and all other) elements contain a nucleus which sits inside orbiting electrons. The nucleus is positively charged and contains particles called protons and neutrons. The electrons are negatively charged. The number of protons in the nucleus determines the type of element.

How Atoms May Vary

If the number of electrons or neutrons in an atom changes it will change the overall charge of the atom and that will affect the chemical behavior of the atom. Such a change will also result in us calling it an ion or an isotope.

Electrons and Their Behavior

Electrons are found in areas around the nucleus called shells. Each shell can be subdivided into two or more orbitals. Electrons must gain energy in order to move to an outer shell. If an electron loses energy it moves to an inner (closer to the nucleus) shell. Each shell can only hold a certain number of electrons, so that any additional electrons must be located in an outer shell. Electrons may move from one atom to another in the processes of oxidation and reduction. Atoms tend to give up or accept additional electrons until the outer electron shell of the atom is filled with electrons.

Chemical Bonding

When atoms join to form molecules they are held together by one of the types of chemical bonds. When a positively charged ion and a negatively charged ion are attracted to each other they are held together by an ionic bond. Covalent bonds are formed when atoms share electrons in their outer electron shells. Hydrogen bonds hold polar (charged) molecules together. Hydrogen bonds are responsible for many of the important characteristics of water which make water so valuable to life as we know it on Earth.

Chemical Reactions

Chemical reactions occur in predictable ways because of the characteristics of atoms. When chemical bonds are made and broken we say that a chemical reaction occurs. In some reactions energy is given off, and in others energy must be added before the reaction takes place. The amount of energy required to start a reaction may be lowered by the addition of a catalyst. In living systems enzymes are important catalysts that speed up most reactions.

The Molecules of Life

Carbon plays a special role in the chemistry of living systems due to the fact that it can share its four outer electrons and form four covalent bonds. Carbon atoms can also form long chains. When functional groups are added to these chains the molecules take on specific characteristics. Carbon chains and functional groups form the basis of the four groups of molecules that make up living systems: carbohydrates, lipids, proteins, and nucleic acids. It is common to find repeating units of smaller molecules in each of these four groups.

Carbohydrates are important to living systems as sources of energy and as the structural chemicals cellulose and chitin. Lipids are utilized as energy storage chemicals, protective waxes, phospholipids of cell membranes, and steroids. Proteins are made of specific sequences of amino acids, and are used as enzymes, hormones, oxygen carriers, and structural components (to name just a few). Nucleic acids (DNA and RNA) are important in genetics, and will be discussed in more detail in a later chapter.

KEY TERMS

Be sure to make and use flash cards for all of these terms. Suggestions are found in "To The Student" at the beginning of the Study Guide. Page numbers refer to the text.

active site	66
amino acid	78
amino group	71
atom	54
atomic number	55
carbohydrate	47
carboxyl group	71
catalyst	66
cellulose	75
chemical reaction	63
chitin	75
CHNOPS	54
compound	54
covalent bonding	60
dehydration	72
denaturation	82
deoxyribonucleic acid (DNA)	82
disaccharide	74
double bond	60
double helix	84
electron	55
element	54
endergonic	67
energy of activation	65
enzyme	66
exergonic	67
fat	75
fatty acid	75
free radical	71
functional group	71
glycerol	75
glycogen	74
hydration	72
hydrocarbon	70
hydrogen bond	61
hydrolysis	72
hydroxyl group	71
inert	58
ion	56
ionic bond	59
isotope	55
kinetic energy	65
lipid	75
molecule	54
monomers	72
monosaccharide	74
neutron	55
nitrogenous bases	82
nucleic acids	82
nucleotide	82
nucleus	55
orbital	56
oxidation	57
peptide	80
peptide bond	80
phospholipid	77
polarized	60
polymer	72
polypeptide	80
polysaccharide	74
primary structure	80
product	67
protein	66
proton	55
quaternary structure	82
reactant	64
reduction	57
ribonucleic acid (RNA)	82
saturated	76
secondary structure	80
shell	56
starch	74
steroid	77
substrate	67
sugar	74
tertiary structure	80
unsaturated	76
wax	77

EXERCISES

Check your understanding of Chapter 3 by completing the following exercises. Answers begin on p. 207.

1. Draw an atom of oxygen in the space below and label the parts.

2. Draw a diagram of a molecule of water in the space below. Label the parts, the bonds, and the positive and negative ends.

True or False Questions

Mark each question either T (True) or F (False).

____3. When an electron gains energy it may move to a higher (outer) shell.

____4. Covalent bonds involve shared electrons.

____5. Proteins are made of sugar monomers.

____6. Enzymes are used to slow down chemical reactions.

Matching Questions

Each question may have more than one answer; write all of the answers that apply. Answers may be used more than once.

____7. the contents of jar full of just neon

____8. a drop of pure water

____9. a lump of sugar

____10. O_2 atoms

a. atoms
b. an element
c. molecule(s)
d. compound(s)

Multiple-Choice Questions

Circle the choice that is the best answer for each question.

For Questions #11-13, assume that you have an atom with six electrons, six neutrons, and six protons.

11. This atom is
    a. positively charged.
    b. negatively charged.
    c. electrically balanced (neutral).
    d, it is not possible to tell from the data given.

12. The atom is called
    a. hydrogen.
    b. carbon.
    c. water.
    d. 6H.
    e. it is not possible to tell from the data given.

13. If you added a proton to this atom it would become
    a. the isotope Carbon 6.
    b. the ion Carbon 6.
    c. the isotope Carbon 13.
    d. the ion Carbon 13.

14. An atom with seven electrons in its outer shell would probably form a chemical bond by
    a. gaining another electron.
    b. giving up an outer electron.
    c. giving up all seven outer electrons.
    d. giving up three electrons to become an ion, then sharing the four electrons that are left.

15. An atom of carbon can form
    a. four ionic bonds.
    b. four covalent bonds.
    c. four hydrogen bonds.
    d. two ionic bonds combined with two covalent bonds.

Fill in the Blanks

Complete each statement by writing the correct word or words in every blank.

16. A single water molecule is held together by __________ bonds, but two different water molecules are held close together by a __________ bond.

17. An atom can have two __________ in the first electron shell and __________ __________ in the second shell.

18. A nucleic acid is a long chain, which could be called a _________. It is made of subunits called _________. Examples of nucleic acids are _________.

Questions For Further Thought

Write your answer for each question in the space provided.

19. What does CHNOPS represent, and why is that important for this course?

20. What allows an enzyme to be so specific that it only catalyzes certain kinds of reactions?

SUMMARY

Study the fourteen items listed in the SUMMARY on p. 84 of the text. You should realize that all the parts of all living things are made up of various atoms, most of which are in molecules. The chemicals of life bond together in predictable patterns and undergo specific reactions, all of which form the basis of an organism's structure and function.

Our knowledge of the makeup and behavior of these atoms and molecules allows us to have a more complete understanding of life and the activities of living things.

HINTS

Read these suggestions before you check your answers to the EXERCISES.

For question 1: See Figure 3.3 and Table 3.2 in the text.

For question 5: This asks about proteins, not carbohydrates.

For question 7: This question may have more than one answer, so write all of the correct answers.

For question 9: Sugar is a carbohydrate.

For question 13: You are adding protons, not electrons.

For question 14: If you formed an ion (as in choice d), it would make an *ionic* bond.

CONGRATULATIONS . . . YOU HAVE COMPLETED CHAPTER 3 !!!

Part Two Cells and Inheritance

CHAPTER 4

THE CELL AND ITS STRUCTURE

CONNECTIONS

Part Two of the text begins with this chapter and focuses on cells and their role in the survival and reproduction of living things. Current cell theory states that all living things are made of cells. When you become familiar with what individual cells are capable of doing you will have a much better understanding of why a large multicellular being such as yourself can and cannot do certain things.

Look for the historical account of how our knowledge of cells developed from the 1600s to the present. In this chapter you will also have your first chance to apply the chemistry that you learned in Chapter Three.

OBJECTIVES FOR THIS CHAPTER

When you have finished Chapter 4, you should be able to:

1. List the components of the cell theory.
2. Describe the development of cytology as a science.
3. Describe the differences between eukaryotic and prokaryotic cells.
4. Explain the advantages of specialization in eukaryotic cells.
5. Describe the components and physical properties of cell walls.
6. Describe the components and physical properties of plasma membranes.
7. Describe the components and physical properties of the cytoskeleton.
8. List the cellular organelles, and describe the functions of each.
9. Show how the fluid mosaic model explains the structure of the plasma membrane.

10. Describe the types of active and passive transport associated with cells.
11. Explain how active and passive transport function to move substances in biological systems.

CHAPTER 4 OUTLINE

I. Cytology and Technology
II. Prokaryotic and Eukaryotic Cells
    A. Specialization in Eukaryotic Cells
III. Cell Components
    A. Cell Walls
    B. Plasma Membrane
    C. The Cytoskeleton
    D. Microtubules
    E. Centrioles
    F. Cilia and Flagella
    G. Mitochondria
    H. Ribosomes
    I. The Endoplasmic Reticulum
    J. Golgi Bodies
    K. Lysosomes
    L. Plastids
    M. Vacuoles
    N. The Nucleus
IV. How Molecules Move
    A. Passive Transport
        1. Diffusion
        2. Facilitated Diffusion
        3. Osmosis
    B. Active Transport
        1. Facilitated Active Transport
        2. Endocytosis and Exocytosis

CONCEPTS IN BRIEF

Cytology and Technology

Light microscopes have allowed scientists to discover a tremendous amount of information about the nature of cells. New technologies in the 1900s have resulted in the development of electron microscopes. All of these tools are used together by modern biologists to study cells. Since the smallest organisms are made of only one cell, and you are a highly organized collection of billions of cells, understanding the structure and function of cells is an important task.

Prokaryotic and Eukaryotic Cells

Bacteria and cyanobacteria are made of prokaryotic cells. All other living things on Earth are made of eukaryotic cells. This course emphasizes eukaryotic organisms and their specialized cells. Cells are more likely to survive changes in their environment if they can control their own activities. Being specialized and remaining small are two ways that they are able to gain more control.

Cell Components

Cells are very active and contain a wide variety of organelles (parts) which are responsible for all cellular activities. There are many different kinds of cells. They are different because they don't all contain the same types and numbers of organelles. Even though organelles may move around, if you know *where* an organelle is you can predict what will take place in that part of the cell. Cell functions are responsible for all of the activities of both single celled and multicellular organisms.

Carbohydrates, proteins, lipids, and nucleic acids are manufactured in cells. These chemicals are then used for making cell structures and regulating cellular activities. Keep in mind that you are alive and functioning because all of this is taking place inside of individual cells throughout your body as you read this.

Cells, and many of their organelles, contain one or two layers of membranes which may even merge with the membranes of other structures. Membranes play many important roles in the life of a cell. Some organelles (mitochondria and chloroplasts) may have originated as free-living organisms that are now integral parts of the cell.

How Molecules Move

Cells must move materials in and out through their membranes, aa well as from one location to another within the interior of the cell. The processes of active and passive transport are responsible for such movement. You may remember that molecules are always in motion. The energy of this motion powers passive transport and results in a *net* movement of molecules from the spot where they are most highly concentrated to the areas of their lowest concentration.

If nothing gets in the way, this net movement (diffusion) will continue until there is an even concentration throughout the entire area. When this state of equilibrium is reached, movement continues, but only in random directions. Several factors can effect the rate of the movement of molecules.

Two special cases of passive transport are included here. First, facilitated diffusion results in carrier proteins assisting certain molecules as they move through membranes. Secondly, osmosis describes the movement of water across membranes. Both of these still show a net movement of the material from *its* region of high concentration to *its* region of low concentration.

In order to stay alive, cells must also pack materials into areas where that material is already in high concentration. This requires that the cell spend some of its own energy, and is called active transport. One method by which this may be accomplished is by carrier molecules which "pump" the material across a membrane in facilitated active transport. A second method occurs when cells acquire (endocytosis) or export (exocytosis) a larger bulk of materials enclosed in packets surrounded by membrane.

KEY TERMS

Be sure to make and use flash cards for all of these terms. Suggestions are found in "To The Student" at the beginning of the Study Guide. Page numbers refer to the text.

active transport	110	cilia	101 & 103
cell theory	92	contractile vacuole	109
cell wall	99	cristae	104
centriole	102	cytology	92

EXERCISES

Check your understanding of Chapter 4 by completing the following exercises. Answers begin on p. 207.

1. Draw an animal cell. Include the following components and briefly state the function of each.

   plasma membrane, nucleus, rough ER, lysosome

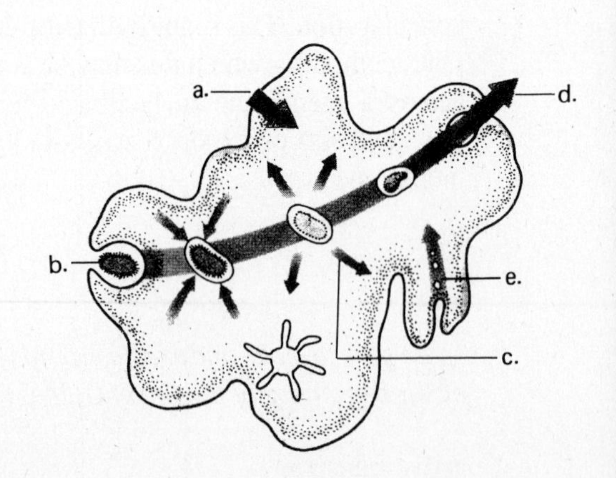

2. Label the proper regions of this diagram with the following terms.

   pinocytosis, exocytosis, phagocytosis, osmosis, diffusion

True or False Questions

Mark each question either T (True) or F (False).

____3. Your body cells are alive.

____4. All cells contain a nucleus.

____5. Osmosis is a form of active transport.

____6. Transmission electron microscopes can distinguish smaller spaces than light microscopes are able to distinguish.

____7. Cell membranes are active regions of the cell.

Matching Questions

Each question may have more than one answer; write all of the answers that apply. Answers may be used more than once.

____8. ribosome

____9. nucleus

____10. mitochondria

a. may once have been free living
b. have double membranes
c. contains DNA
d. site of protein production
e. regulates protein production

____11. found in plant cells but not in animal cells
____12. found in animal cells but not in plant cells
____13. may be found in both plant and animal cells

a. vacuole
b. nucleus
c. cell wall
d. plasma membrane
e. plastid
f. mitochondria
g. all of the above
h. none of the above

Multiple-Choice Questions

Circle the choice that is the best answer for each question.

14. The plasma membrane
    a. gives the cell control over its internal environment.
    b. regulates what passes into the cell.
    c. protects the cell from the external environment.
    d. regulates what passes out of the cell.
    e. all of the above.

15. Specialized cellular organelles
    a. make it more difficult for the cell to function efficiently.
    b. are not found in eukaryotic cells.
    c. give the cell an advantage it would not have without specialization.

d. show that the endosymbiotic theory is no longer acceptable.
e. all of the above.

16. Plasma membranes
    a. are flexible.
    b. regulate what goes in and out of cells.
    c. surround the fluid part of the cell.
    d. may have parts of the cytoskeleton attached to them.
    e. all of the above.

17. The cytoskeleton of the cell
    a. holds organelles in place.
    b. makes the cell too rigid to move around in cold weather.
    c. is responsible for facilitated active transport.
    d. all of the above.

18. Cells move from place to place by the action of
    a. plastids.
    b. lysosomes.
    c. cilia and flagella.
    d. rough and smooth ER.
    e. all of the above.

19. The cell theory was first developed
    a. in the 1500s.
    b. in the 1800s.
    c. after the development of the electron microscope.
    d. as a result of the endosymbiotic theory.

Fill in the Blanks

Complete each statement by writing the correct word or words in every blank.

20. Cells come from other __________.

21. When plant cells are surrounded by water, the process of __________ will cause them to fill with water. Their __________ prevents them from bursting.

22. The fluid mosaic model states that plasma membranes are primarily composed of a double layer of __________ molecules. There are __________ molecules imbedded in them. __________ molecules are also present and seem to give the cell type its identity.

Questions For Further Thought

Write your answer for each question in the space provided.

23. Mitochondria are [more, less] (circle one) common in active types of cells than in less active ones. Explain why.

24. Discuss how the development of electron microscopes helped the study of cytology.

SUMMARY

Study the eighteen items listed in the SUMMARY on p. 116 of the text. By now it should be clear that what an organism does is a reflection of what its cells are doing. Technological advances have been, and will continue to be, very important in cytology. It is clear that cells are very active and that they contain many complex parts. An understanding of the chemical (molecular) structure of these cellular components helps us explain the functions of each part of the cell. This knowledge has also been useful for theorizing about the history and development of modern cells.

HINTS

Read these suggestions before you check your answers to the EXERCISES.

For question 4: Consider both prokaryotic and eukaryotic cells.

For question 6: Don't forget about resolution.

For question 8: Note that choices d and e are *not* the same.

For question 10: Remember the endosymbiotic theory?!?

For question 19: This does not ask when cells were first seen.

For question 21: This assumes passive transport.

For question 23: Remember that mitochondria are associated with energy reactions.

CONGRATULATIONS . . . YOU HAVE COMPLETED CHAPTER 4 !!!

CHAPTER 5

ENERGY: THE DANCE OF LIFE

CONNECTIONS

By this point in the course you have learned that cells are the functional units of living things and that cellular chemistry determines what goes on in an organism. Now it is time to focus on the activities of two organelles, the chloroplasts and mitochondria, and their effects on cellular energy supplies.

Energy is required for all movement. In order for cells (and, of course, larger organisms) to repair themselves, move, grow, and reproduce they must have constant energy supplies. This chapter describes the characteristics of energy and the chemical reactions associated with energy exchanges inside cells. Watch for common features shared by plants and animals in the ways they acquire, store, and use energy supplies.

OBJECTIVES FOR THIS CHAPTER

When you have finished Chapter 5, you should be able to:

1. Define the two states of energy.
2. List several forms of energy.
3. State the laws of thermodynamics.
4. Explain the implications of the laws of thermodynamics for living systems.
5. Describe how the ATP molecule is formed.
6. Explain the role that ATP plays in supplying energy for cells.
7. Summarize the photosynthetic process by tracing the steps in the conversion of light energy to chemical energy.
8. Relate the structure of a chloroplast to its function.
9. Describe the four major processes in cellular respiration, stating where each takes place and listing the materials used and the products.
10. Compare and contrast the role of the electron transport system in the chloroplast and the mitochondrion.

CHAPTER 5 OUTLINE

- I. The Laws of Thermodynamics
- II. ATP: The Energy Currency
  - A. Photosynthesis: A Poet's View
  - B. Photosynthesis: A Closer View
    1. The Light-Dependent Reactions
    2. The Light-Independent Reactions
    3. What Does a Plant Do with Its Glucose?
- III. Food to Energy
  - A. Food to Energy: A Poet's View
  - B. Food to Energy: A Closer View
    1. Glycolysis
    2. Fermentation
    3. Aerobic Respiration
    4. The Krebs Cycle
    5. The Electron-Transport System
    6. Chemiosmotic Phosphorylation
- IV. A Brief Review

CONCEPTS IN BRIEF

The Laws of Thermodynamics

The laws of thermodynamics describe what happens to energy as it is used in both living or nonliving systems. The effects of these laws cannot be avoided. Because of the first law of thermodynamics we encounter two important effects: (1) plants are able to convert the energy of the sun into sugars, and (2) both plants and animals are able to convert sugars into the energy they need in order to move and function. Due to the second law of thermodynamics, however, all organisms require a constant source of new energy to replace what they have used, and therefore, converted into heat.

ATP: The Energy Currency

While you are reading this study guide your cells are using the energy from the bonds of ATP molecules for their activities. In fact, ATP is used to provide the energy needed by all living things. Originally that energy came from the sun and was converted into the bond energy of sugar molecules by green plants during the process of photosynthesis. These conversions of light energy into usable cellular energy involve the moving of electrons from molecule to molecule during numerous chemical reactions inside chloroplasts.

During the first stage of photosynthesis, known as the light-dependent reactions, light energy strikes the chlorophyll molecules of both photosystem II and photosystem I. This energizes and liberates electrons. In addition, water molecules are broken apart and the oxygen from the water is released into the environment. The hydrogen electrons from the water are captured by the chlorophyll of photosystem II, while the hydrogen ions (protons) are moved across membranes and used during chemiosmosis to generate ATP.

The energized electrons from photosystem II are captured by the chlorophyll of photosystem I, and the electrons that were released from photosystem I are used to generate the energy-rich NADPH. Notice that all of the parts are accounted for, and the chloroplast is left with energy stored in two kinds of high-energy molecules: ATP and NADPH.

The next stage, known as the light-independent reactions, uses the recently created ATP and NADPH for energy rather than using light. It is here that the CO_2, which has been taken into the plant from the environment, is added to existing carbon-containing molecules in the Calvin cycle. The end result is generally the production of glucose from a three-carbon compound called PGAL. This glucose may be used right away or it may be stored as starch for future use.

Food to Energy

All living things must use cellular respiration and convert the energy in their foods (commonly glucose by the time it reaches the cells) into the energy of ATP. The stages of respiration may be aerobic (using oxygen) or anaerobic (not using oxygen). The four major sets or reactions to look at are: glycolysis, the Krebs cycle, electron transport, and chemiosmotic phosphorylation.

The first stage, glycolysis, produces a net gain of two ATP molecules and leaves the cell with two molecules of pyruvate (also called pyruvic acid). In aerobic organisms that have oxygen present, the pyruvate is converted into acetyl-CoA, which enters the Krebs cycle. However, if the organism is anaerobic, or if it is an oxygen-starved aerobic organism, the pyruvate goes through fermentation reactions.

When acetyl-CoA enters the Krebs cycle in the mitochondrion it is oxidized. Electrons, H^+ protons, and CO_2 are released. Energy from the liberated electrons is used to generate the energy-rich molecules $FADH_2$, NADH, and ATP, while the CO_2 is thrown away as a waste product.

The newly acquired electrons of the $FADH_2$ and NADH molecules are then passed to other electron acceptors of the electron transport system and their energy is slowly released. This energy is used to move the previously released hydrogen ions through membranes (during chemiosmotic phosphorylation) and finally, a large number of ATP molecules is produced. If you are still wondering where the *aerobic* aspect comes in, look at the end of the electron transport chain. You should see oxygen (which came into the organism from the environment) acting as an electron acceptor and forming water.

A Brief Review

As you can guess from the title, this section reviews photosynthesis and respiration. Note that in each of these processes there is an electron transport system in which hydrogen ions move across a membrane, providing energy that powers the conversion of ADP + P_i to ATP. The ATP produced in photosynthesis is used during the formation of PGAL. The gain in ATP during respiration reactions provides the organism with high energy molecules which can be used for all its metabolic needs.

KEY TERMS

Be sure to make and use flash cards for all of these terms. Suggestions are found in "To The Student" at the beginning of the Study Guide. Page numbers refer to the text.

acetate	142	chemiosmosis	131
acetyl-CoA	142	chemiosmotic phosphorylation	146
adenosine diphosphate (ADP)	124	chlorophyll	127
adenosine triphosphate (ATP)	124	chlorophyll a	127
alcoholic fermentation	140	chlorophyll b	127
anaerobic	136	citrate	143
Calvin cycle	131	electrical energy	123
carotenoids	127	electromagnetic spectrum	130
cellular respiration	133		

EXERCISES

Check your understanding of Chapter 5 by completing the following exercises. Answers begin on p. 207.

1. Arrange the following events of photosynthesis in sequence from first to last.
   a. Free oxygen is formed from water.
   b. Light hits the plant.
   c. Glucose is formed.
   d. CO_2 is joined to RuBP.
   e. NADPH is formed.
   f. PGAL is formed.

____ ____ ____ ____ ____ ____

First Last

2. Arrange the following events of aerobic respiration in sequence from first to last.
   a. CO_2 is formed.
   b. H_2O is formed.
   c. Glucose enters the cell.
   d. $NFADH_2$ is used.
   e. Pyruvate is formed.
   f. Acetyl-CoA is formed.

____ ____ ____ ____ ____ ____

First Last

3. This diagram represents part of a chloroplast. Label the indicated areas with the following terms: thylakoid, stroma, photosystem I and photosystem II, accumulating protons, glucose formation, membranes

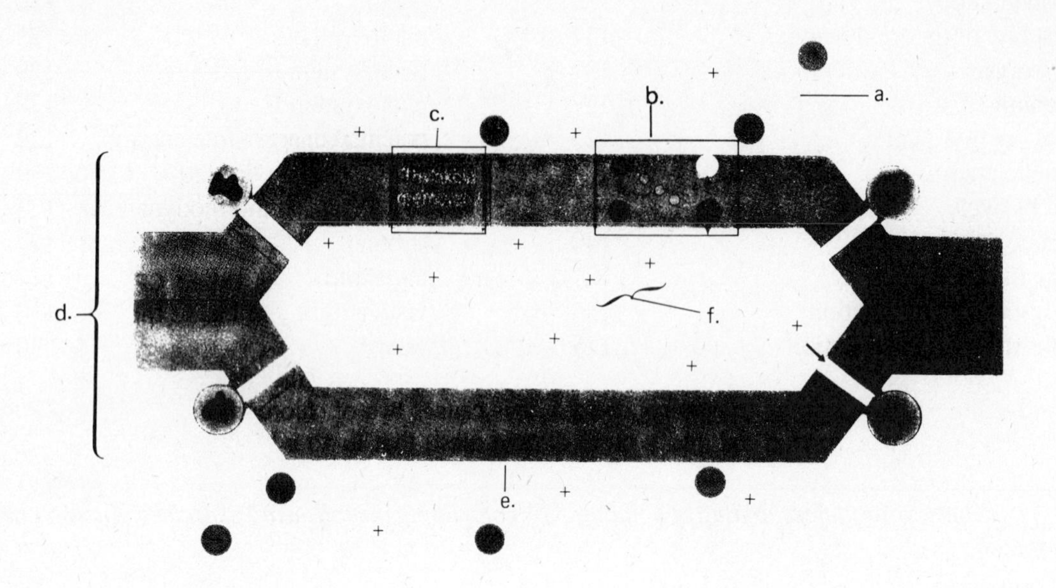

True or False Questions

Mark each question either T (True) or F (False).

____4. Oxygen is needed for the light-dependent reactions to take place.

____5. An electron transport system is found in both the mitochondria and in chloroplasts.

____6. Chlorophyll is necessary for photosynthesis but not for respiration.

____7. Your body cells are aerobic.

Matching Questions

Each question may have more than one answer; write all of the answers that apply. Answers may be used more than once.

____8. uses energy from NADHs

____9. the energy released is used to pump hydrogen ions across a membrane

____10. as electrons are passed along they gradually lose energy

____11. electrons came from chlorophyll molecules

a. electron transport system of photosynthesis only
b. electron transport system of respiration only
c. electron transport system of both photosynthesis and respiration

____12. produces the most ATP.

____13. converts glucose to pyruvate.

____14. gives off CO_2.

____15. may lead to fermentation reactions.

____16. takes place in the mitochondria.

a. Glycosis

b. Electron transport

c. Krebs cycle

d. Chemiosmotic phosphorylation

Multiple-Choice Questions

Circle the choice that is the best answer for each question.

17. The _____ states that it is possible to convert food energy into the energy of motion.
    a. first law of thermodynamics
    b. second law of thermodynamics
    c. third law of thermodynamics

18. Heat is generated as a result of
    a. all energy conversions.
    b. respiration reactions, but not photosynthesis reactions.
    c. aerobic respiration but not anaerobic respiration.
    d. specific reactions in warm-blooded animals, but not in other organisms.

19. During photosynthesis, light energy itself is used to
    a. join carbons together to form glucose.
    b. turn CO_2 into O_2.
    c. drive the reactions of chemiosmotic phosphorylation.
    d. energize electrons in chlorophyll.
    e. all of the above.

20. When simpler molecules are joined together to form more complex molecules,
    a. energy is stored.
    b. energy is released.

Fill in the Blanks

Complete each statement by writing the correct word or words in every blank.

21. The light-dependent reactions occur in the __________ of the __________ portion of the chloroplast.

22. The Krebs cycle occurs in the organelle called the __________.

23. During chemiosmotic phosphorylation, __________ is formed when __________ ions are pumped across the __________ of the mitochondrion.

24. If four molecules of ATP release their energy, how many molecules of ADP will be formed?
____

25. In the presence of oxygen, aerobic respiration produces a total of __________ molecules of ATP per molecule of glucose. Anaerobic respiration produces a total of __________ molecules of ATP per molecule of glucose.

Questions For Further Thought

Write your answer for each question in the space provided.

26. Explain why, even if you water and fertilize it properly, a green plant will die if it is kept in the dark.

SUMMARY

Study the thirteen items listed in the SUMMARY on p. 147 of the text. It should be clear to you now that *both* photosynthesis and respiration are essential to life on Earth. Photosynthesis allows plants and algae to convert sunlight into a form of chemical energy (food) that is useable by living things for their daily energy needs. Respiration reactions provide all living things with the ability to convert the energy in their food into the energy of ATP molecules which they can use to do work. Hopefully you have noticed the universal nature of these reactions.

HINTS

Read these suggestions before you check your answers to the EXERCISES.

For question 4: This is photosynthesis, not respiration.

For question 12: The key word here is "most."

For question 16.: The key word here is "in."

For question 18: The second law of thermodynamics applies to all living and nonliving systems.

For question 19: This deals with what light does directly, not indirectly.

For question 26: All living things require energy such as glucose.

CONGRATULATIONS . . . YOU HAVE COMPLETED CHAPTER 5 !!!

CHAPTER 6

THE CELL AND ITS CYCLES

CONNECTIONS

Now it is time for a more thorough examination of the nucleic acids (RNA and DNA) which Chapter Three introduced to you. The current chapter presents the details of how RNA and DNA can be responsible for regulating the daily operations and reproduction of cells, and consequently, all organisms. If you keep thinking about the important role of cellular chemistry, this topic will be easier to understand.

Genetics is coming up soon, in fact, in the next chapter. Chapter Six, however, presents the background that you need for understanding how and why you look like (but not *exactly* like) other members of your family. Notice that this chapter looks at both how cells survive and how they reproduce.

OBJECTIVES FOR THIS CHAPTER

When you have finished Chapter 6, you should be able to:

1. Explain the overall function of the cell cycle.
2. List the sequence of events that occur during the stages of the cell cycle and describe the major events in each.
3. List the stages of meiosis in sequence and describe what happens to the chromosomes during each stage.
4. Explain how meiotic events halve chromosome numbers during gamete formation.
5. Compare and contrast the stages and end results of mitosis and meiosis.
6. Describe the components and spatial arrangement of the DNA molecule as proposed by James Watson and Francis Crick.
7. Explain how the DNA molecule replicates.
8. Relate the structure of DNA to its replication and the synthesis of RNA.
9. Describe how chromosomes direct the production of proteins via transcription and translation.
10. Compare and contrast the properties of RNA and DNA.
11. Explain what is meant by the genetic code.

CHAPTER 6 OUTLINE

I. The Cell Cycle and Mitosis
II. Meiosis
III. The Double Helix
IV. DNA Replication
V. How Chromosomes Work
 A. Transcription
 B. The Genetic Code
 C. Translation
VI. Continuing Problems in Cell Biology

CONCEPTS IN BRIEF

The Cell Cycle and Mitosis

Individual cells have a life which is detailed in the events of the cell cycle. When a cell reproduces by mitosis and divides by cytokinesis it becomes two genetically identical daughter cells. Since the parent cell had already reproduced its genetics, these daughter cells each have the same genetics that the parent cell had before it went through mitosis and cytokinesis. A very active phase (interphase) follows which may lead to mitosis again. The events of these stages follow a certain sequence and are predictable. During much of mitosis the chromosomes are visible with a light microscope and their locations and activities can be observed.

Meiosis

Meiosis involves specialized cells in the reproductive organs of living things which reproduce sexually. These diploid cells reproduce their chromosomes before going through two divisions and becoming four haploid sperm (in males) or one haploid egg and four polar bodies (in females). The haploid sperm and eggs contain just one member of each chromosome pair. This, along with crossing over, provides for increased genetic variation in the offspring of the organisms involved.

Following the chromosomes shows that mitosis and meiosis have some similar features yet are very different. Crossing over only occurs in meiosis, and only during prophase I. In the first division of meiosis each replicated chromosome stays intact (as two chromatids) and the paired (homologous) chromosomes move into different cells. The chromatids of each chromosome do not separate until anaphase II, the second division of meiosis.

The Double Helix

James Watson and Francis Crick received credit for discovering the structure of DNA in the 1950s. The molecular structure of DNA (a double helix) relates closely to its various functions. Two of these important functions are replicating itself, and making molecules of RNA. The nitrogenous bases of DNA always join in specific pairs (A with T, and C with G) in the inner part of the molecule where they are held together by weak hydrogen bonds. The *sequence* of these bases determines the genetic information of the molecule.

DNA Replication

DNA replication begins when the base pairs separate and the two strands "unzip". This allows nucleotides which are free in the nucleus to attach to the exposed bases of *each* strand of the original

DNA molecule. When replication is complete there are two complete DNA molecules which are identical to each other. These two also have the same genetics that the original DNA molecule had before it replicated.

How Chromosomes Work

Cells are different from one another because not all cells produce the same types of proteins. The chromosomes of each cell regulate *which* proteins that cell produces. Therefore, it can be said that the genetic information (the sequence of DNA bases) of the chromosomes accounts for the differences found in different types of cells.

The process of protein production begins with sections of DNA making the various types of RNA during transcription. This is followed by each molecule of mRNA and numerous tRNA molecules meeting at a ribosome (made of rRNA) in the cytoplasm. Here the anticodons of numerous tRNAs match the codons of the mRNA. Amino acids that were carried by the tRNAs are joined into a chain in a specific sequence during translation. Therefore, the sequence of amino acids is determined by the sequence of bases on the mRNA. Table 6.4 in the text can be used to predict the amino acid sequence from any known mRNA sequence. The newly made amino acid chains form polypeptides or proteins which determine the nature of the cell.

Continuing Problems in Cell Biology

The CONNECTIONS section of Chapter 2 in this study guide mentioned that scientists do not have the answers to everything. The current text section reminds you of this in terms of cells and their molecular components. New information on these and other topics is constantly being made available, and there is no end in sight.

KEY TERMS

Be sure to make and use flash cards for all of these terms. Suggestions are found in "To The Student" at the beginning of the Study Guide. Page numbers refer to the text.

adenine	166	gonad	158
anaphase	152 & 153	guanine	166
anticodon	172	haploid	158
cell cycle	152	homologue	157
centromere	153	interphase	152
chromatid	153	meiosis	158
chromosome	153	messenger RNA (mRNA)	169
codon	171	metaphase	152 & 153
crossing over	159	mitosis	152
cytokinesis	152	nondisjunction	165
cytosine	166	parent cell	152
daughter cell	152	polar body	164
deoxyribonucleic acid (DNA)	164	prophase1	52 & 153
diploid	158	purines	166
gamete	158	pyrimidine	166
genetic code	171	ribonucleic acid (RNA)	169

ribosomal RNA (rRNA)	169	transfer RNA (tRNA)	169
spindle	153	transcription	169
telophase	152 & 156	translation	169
thymine	166	uracil	169

EXERCISES

Check your understanding of Chapter 6 by completing the following exercises. Answers begin on p. 207.

1. Look at Figure 6.7 (B) in your text.

   a. How many chromosomes are shown?

   b. How many chromatids are shown?

   c. How many centromeres are shown?

2. a. The following drawings represent four of the stages of [mitosis, meiosis] (choose one).

   b. Number the drawings (1, 2, 3, 4), using 1 for the stage which would come first and 4 for the last stage.

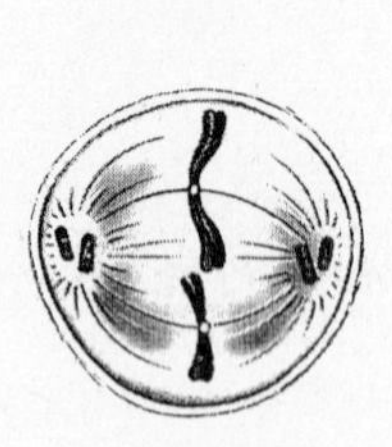
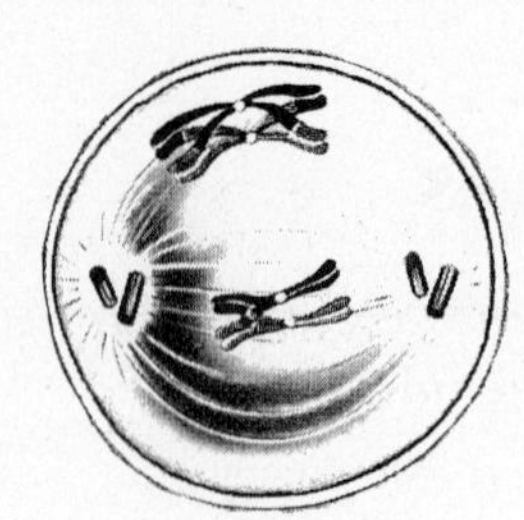
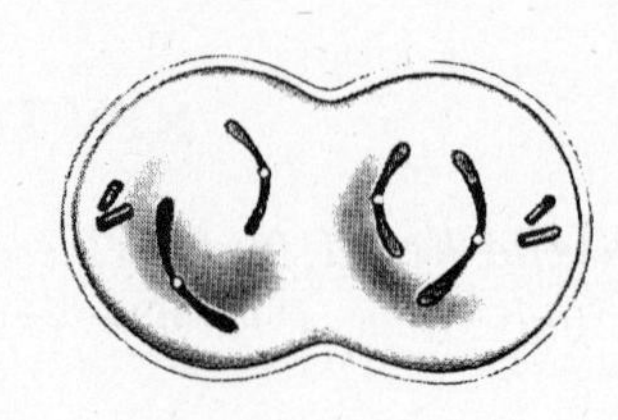

______ ______ ______ ______

3. Assume that you have a strand of DNA with the following sequence of bases: TACGGCTAACGTATT.

   a. What mRNA strand will form from this sequence?

   b. What sequence of amino acids will be formed during protein synthesis?

c. What is the sequence of bases in the *other* strand of DNA?

True or False Questions

Mark each question either T (True) or F (False).

____4. Anticodons are found on mRNA.

____5. Uracil is found in both RNA and DNA.

____6. The cell cycle illustrates that cells come from other cells.

____7. Interphase of mitosis is an inactive or "resting " phase for the cell.

____8. The two divisions during meiosis result in halving the chromosome.

Matching Questions

Each question may have more than one answer; write all of the answers that apply. Answers may be used more than once.

____9. chromosomes replicate

____10. homologous chromosomes lay next to each other as pairs

____11. the cells that result are diploid

____12. takes place throughout most of your body

a. occurs only in mitosis
b. occurs only in meiosis
c. occurs in both mitosis and meiosis

____13. contain(s) adenine

____14. contain(s) uracil

____15. is (are) usually double stranded

____16. contain(s) a "backbone" of sugar and phosphate

____17. new molecules of this are formed in the nucleus

a. DNA only
b. RNA only
c. both DNA and RNA

Multiple-Choice Questions

Circle the choice that is the best answer for each question.

18. If a cell with 18 chromosomes undergoes mitosis, it will produce _____ cells, each of which will have _____ chromosomes.
    a. 2, 9
    b. 2, 18
    c. 4, 9
    d. 4, 18

19. A specific molecule of tRNA can carry __________ kind(s) of amino acid(s).
    a. 1
    b. 3
    c. 20
    d. 64

20. Meiosis allows for __________ genetic variation in the offspring.
    a. an increase in
    b. a decrease in
    c. neutralization of
    d. no effect on

21. Crossing over occurs during
    a. prophase of mitosis.
    b. prophase I of meiosis.
    c. all stages of mitosis.
    d. all stages of meiosis.
    e. both c and d are correct.

Fill in the Blanks

Complete each statement by writing the correct word or words in every blank.

22. When an alligator (see Table 6.1 in the text) makes sperm, the cells of its testes go through the process of __________. Each single cell will produce [1, 2, 4, 8] (choose one) sperm. Each of these sperm will be [*haploid, diploid*] (choose one) and will have [2, 4, 8, 16, 32, 64] (choose one) chromosomes.

23. During protein synthesis, RNA is formed from DNA by the process of __________, and amino acids are sequenced in the process of __________.

24. The genetic code consists of sequence of __________ (write a number) nitrogenous __________.

Questions For Further Thought

Write your answer for each question in the space provided.

25. Explain why it is the *sequence* of bases in DNA which is so critical to protein production.

26. Since essentially all of your body cells have the same genetics, explain why you have different *types* of cells in your body.

SUMMARY

Study the eleven items listed in the SUMMARY on p. 177 of the text. You have seen that mitosis provides copies of cells for growth, repair, and cell replacement. Meiosis, on the other hand, provides the sex cells (mostly eggs and sperm) which are used for producing new individual organisms. Although there are similarities in these two processes they yield very different results.

In addition, it should now be clear that it is the *sequence of bases* in the DNA molecules of your chromosomes which is ultimately responsible for determining your characteristics and operating your body. All of this control is coming from the nuclei of your cells.

HINTS

Read these suggestions before you check your answers to the EXERCISES.

For question 1:	Don't forget that these are replicated chromosomes.
For question 3:	Remember that the bases in RNA and DNA are not all the same. Also, be sure to use the *mRNA* strand when you read Table 6.4 in the text.
For question 7:	This concerns *current* knowledge.
For question 12:	Think about where meiosis takes place.
For question 18:	This is *mitosis.*
For question 19:	This is concerned with just one type of tRNA.
For question 22:	The first answer is either mitosis or meiosis.
For question 26:	This is not the same as question . All of your body cells have the *same* nucleotide sequence yet they are specialized into different types.

CONGRATULATIONS . . . YOU HAVE COMPLETED CHAPTER 6 !!!

CHAPTER 7

INHERITANCE: FROM MENDEL TO MOLECULES

CONNECTIONS

You are returning to the 1800s, the time of Darwin. What *you* have already learned about chromosomes in this course was not yet known, and the mystery of how traits were passed from parents to offspring was unsolved. This chapter of your text shows you the contributions and impact of Gregor Mendel's principles of genetics and how they have been built upon by other researchers. In addition, you will be challenged to consider the effects of current and future technological advances that are a part of the very active field of genetic engineering.

OBJECTIVES FOR THIS CHAPTER

When you have finished Chapter 7, you should be able to:

1. Describe the features that made Mendel's work unique.
2. Explain Mendel's three principles and fully describe how his experimental crosses illustrate each.
3. Use a Punnett square to predict phenotypic and genotypic ratios of genetic crosses.
4. Explain the function and effects of a testcross.
5. Relate the behavior of chromosomes during meiosis to Mendel's principles.
6. Describe how sex chromosomes determine gender and list some human sex-linked traits.
7. Show why sex-linked traits result in different genetic ratios than traits on autosomes.
8. Describe the technique of chromosome mapping.
9. Distinguish between gene mutations and chromosome mutations.
10. List the types of mutations that occur at the gene and chromosomal levels.
11. Describe the mechanisms that account for deviations in the ratios predicted by Mendel's principles.
12. State how the Hardy-Weinberg principle accounts for constant allele frequencies within a population.

CHAPTER 7 OUTLINE

I. Mendelian Genetics
  A. A Kindly Old Abbot Puttering in His Garden
  B. The Principle of Dominance
  C. The Principle of Segregation
  D. The Principle of Independent Assortment
  E. The Testcross
II. Classical Genetics - Beyond Mendel
  A. Sex Determination
  B. Gene Linkage and Crossover
    1. Sex-linked traits
III. Chomosome Mapping
IV. Mutations
  A. Gene Mutations
  B. Chromosome Mutations
V. Further Beyond Mendel
  A. Dominance Relationships
  B. Multiple Alleles
  C. Gene Interaction
  D. Polygenic Inheritance
  E. Other Sources of Variation
VI. Population Genetics

CONCEPTS IN BRIEF

Mendelian Genetics

Gregor Mendel was a monk as well as being a bright, well-trained scientist who began to apply mathematics to biological concepts in the 1850s. Through careful planning and patient research on true-breeding garden peas, he was able to formulate three principles of genetics that provide the basis of our current knowledge in that field.

Mendel's principles of dominance, segregation, and independent assortment deal with which traits are passed on by a particular set of parents, and which ones show up in the offspring. He realized that traits are controlled by two factors (now called genes) and that offspring must have received one factor of the pair from each of their parents. His principle of segregation states that when eggs or sperm are formed each of these new cells will carry only one allele of the pair, not both. Remember that *you* know meiosis produces haploid cells.

Once an egg is fertilized it is diploid and has an allele from each parent for every trait. Mendel found that if both alleles are not alike (a heterozygous genotype) usually just one of them will be expressed (the dominant trait) and one will not show up (the recessive trait). This is his principle of dominance.

Today we use capital letters to represent the dominant trait and lower case letters to represent the recessive trait. Making a Punnett square is a convenient method of predicting the possible genotypes and phenotypes of offspring from a mating between two parents. Remember that a Punnett square gives the results that you would expect from a large number of offspring, and is not a guarantee. When you construct a Punnett square, the letters (alleles) on the outside of the square represent all the possible gametes which the offspring could receive. Write the letters inside the boxes by filling in the letters for the gamete allele in all the boxes in that row.

When Mendel followed two characteristics at a time he discovered that two different traits don't always follow each other around. Another method of describing this is to say that the way in which one pair of alleles segregates does not depend on the way another pair of alleles segregates (remember anaphase I?). This became his principle of independent assortment.

It is possible to determine the genotype of an organism that shows a dominant phenotype by doing a test cross. You must cross it with another organism which has a recessive phenotype and examine the phenotypes of their offspring.

Classical Genetics — Beyond Mendel

Once Mendel's work was rediscovered, meiosis was described in detail, and the chromosome theory of inheritance was established, scientists continued to make advances in genetics. The use of the fruit fly, *Drosophila melanogaster*, for genetic research has allowed many studies on mutations, inheritance patterns, gene loci, linkage, and crossing over to take place in short periods of time.

Sex chromosomes determine the gender (XX is female and XY is male) of fruit flies and many mammals. They also carry information about other traits, such as red-green color blindness, hemophilia, and a type of muscular dystrophy, all of which are found on the X chromosome but not on the Y chromosome.

The alleles which are found on a single chromosome are said to be linked. Unless crossing over occurs, these alleles will travel through meiosis as a unit. When crossing over occurs, however, paired chromosomes exchange parts of themselves with each other and alleles which were linked on the same chromosome are now on different chromosomes. This, of course, leads to increased variation in the offspring that is not predicted by a Punnett square.

Chromosome Mapping

The specific location of each gene on a chromosome is found through chromosome mapping. If genes are found far apart from each other, they are more frequently involved in crossing over. However, if they are located nearer to one another, they will not cross over as often. An entire chromosome map can be constructed by examining the frequencies of crossing over of the genes.

Mutations

A change in an organism's hereditary material is called a mutation. Mutations may occur in both genes and chromosomes. If the mutation involves an egg or sperm, the change may show up in the offspring. Both gene and chromosome mutations may be very serious or they may have no effect at all, depending on how the nucleotides were altered. Although some mutations are beneficial to the organism, it is more common for mutations to be harmful.

Further Beyond Mendel

It is now clear that there are many types of inheritance in addition to the simple dominance described by Mendel. Many traits are governed by incomplete dominance in which a heterozygous genotype gives a phenotype that seems to be a mix of the traits of the parents. Instead of only two kinds of alleles, traits such as human blood type are controlled by three or more different alleles. In these cases (called multiple alleles) the alleles still occur in pairs, there are just more possible kinds of pairs.

Polygenic inheritance occurs when more that one pair of alleles interacts to produce a phenotype. Some genes may actually turn other genes off, as in epistasis. In addition, the effects of genes may be influenced by the sex or age of the organism and by factors in the environment.

Population Genetics

It is important to look beyond the genetics of individuals and examine the total set of alleles in the population. The proportion of dominant and recessive alleles (the allele frequencies) in the entire population will remain the same from one generation to another as long as certain conditions are met. Although it is common for these conditions to be violated under natural conditions, they provide scientists with a basis for looking at what causes changes in the genetics of populations over time. A change in allele frequencies is known as evolution.

KEY TERMS

Be sure to make and use flash cards for all of these terms. Suggestions are found in "To The Student" at the beginning of the Study Guide. Page numbers refer to the text.

Term	Page
allele	186
autosomes	192
Barr body	192
chromosome mutation	199
classical genetics	189
complete dominance	199
deletion	199
dihybrid	187
dominant trait	184
duplication	199
epistasis	200
gene	186
gene linkage	193
gene mutation	197
genotype	186
heterozygous	187
homozygous	186
incomplete dominance	199
inversion	199
locus	186
multiple alleles	200
mutation	190
phenotype	186
polygenic inheritance	201
population	202
principle of dominance	184
principle of independent assortment	187
principle of segregation	186
Punnett square	187
recessive trait	184
sex chromosomes	191
sex-linked trait	195
testcross	188

EXERCISES

Check your understanding of Chapter 7 by completing the following exercises. Answers begin on p. 207.

1. The following diagram is a simple chromosome map. The letters represent alleles.

   a. Which allele(s) will show the effects of crossing over the most frequently? __________

b. Which allele(s) will show the effects of crossing over the least frequently? __________

a b c d e f g

c. Will "a" normally cross over with "g"? _____ Explain.

True or False Questions

Mark each question either T (True) or F (False).

_____2. If a gene mutation does not result in a change in the amino acids produced, the phenotype will still be changed.

_____3. If there are no mutation effects and no crossing over, a cross that follows one trait will give a 3:1 ratio of offspring.

_____4. Most mutations end up being helpful to the survival of the organism.

_____5. In time, most members of a population will have a dominant phenotype and few, if any will show a recessive phenotype.

_____6. A Punnett square can be used to predict genotype ratios but not phenotype ratios.

_____7. The Hardy-Weinberg principle states that allele frequencies will remain the same in populations over time.

Matching Questions

Each question may have more than one answer; write all of the answers that apply. Answers may be used more than once.

_____8. Flower color in snapdragons is an example of incomplete dominance. If a red-flowered plant is crossed with a white-flowered plant, what offspring will result?

_____9. What about white crossed with white?

_____10. What about red crossed with red?

a. all red
b. all white
c. half are red and half are white
d. 3 red : 1 white
e. all pink
f. 1 red : 1 pink : 1 white
g. 3 red : 1 pink : 1 white
h. cannot tell from the data given

____11. What gametes could be produced by genotype HhTT?
____12. What gametes could be produced by genotype HHTT?
____13. What gametes could be produced by genotype hhTt?
____14. What gametes could be produced by genotype HhTt?

a. HT
b. Ht
c. hT
d. ht

Multiple-Choice Questions

Circle the choice that is the best answer for each question.

15. If you knew that simple dominance was working, an allele represented by the letter B would be
    a. a dominant allele.
    b. a recessive allele.
    c. either a dominant or a recessive allele, depending on the organism.
    d. stronger than C but not as strong as A.

16. Assume that you crossed a true-breeding garden pea which has green seed pods (G) and one which has yellow seed pods (g). What genotypes would the F_1 generation have?
    a. 3 GG to 1 gg
    b. half GG and half gg
    c. all GG
    d. all Gg
    e. all gg

17. In humans, a boy baby gets his X chromosome from
    a. his father.
    b. his mother.
    c. either his father or his mother.
    d. accumulations of several Barr bodies.

18. Which of the following relates to Mendel's principle of segregation?
    a. Dominant traits and recessive traits separate in meiosis.
    b. Paired alleles end up in different gametes.
    c. Sex determination is not affected by autosomes.
    d. Gene mutations are different from chromosome mutations.
    e. All of the above.

19. The principle of independent assortment affects crosses such as
    a. Rr X Rr
    b. RrTt X RrTt
    c. both of the above
    d. none of the above

Fill in the Blanks

Complete each statement by writing the correct word or words in every blank.

20. In order to determine the genotypes of the green-podded F_1 plants in question , you could do a

    __________ cross. This involves crossing the green-podded F_1 plants with plants that have

__________ pods. You would know if the green-podded ones are heterozygous, with genotype

__________, if any of the F_2 plants show the phenotype __________.

21. A change in allele frequencies is called __________.

22. What an organism looks like is its __________.

23. Human blood types are an example of a trait that is governed by __________ alleles. The alleles involved are _____, _____, and _____.

Questions For Further Thought

Write your answer for each question in the space provided.

24. List some of the factors that can result in ratios of offspring that are different than would be predicted by Mendel's principles alone.

25. Why don't allele frequencies always remain the same in natural populations?

SUMMARY

Study the sixteen items listed in the SUMMARY on p. 204 of the text. This entire chapter reminds you that an organism's characteristics are controlled by its genetics. It should also be clear to you now that there are specific ways in which alleles are passed from parents to offspring. However, there are many *different* methods for this, and there are a wide variety of ways in which the expected results can be altered in all situations. This is another reminder that living systems can appear to be simple, yet are generally quite complex as well. All of the variety may seem confusing but it certainly keeps biology interesting!

HINTS

Read these suggestions <u>before</u> you check your answers to the EXERCISES.

For question 2:	Don't forget everything you learned about protein synthesis.
For question 3:	What about something like rr X rr?
For question 5:	Remember the Hardy-Weinberg principle!?!
For question 7:	Are there any conditions that must be met?
For question 8:	This is *incomplete* dominance, *not* simple dominance.
For question 11:	Write all the possible choices. Assume no linkage and no crossing over.
For question 16:	This is an example of simple dominance
For question 18:	Choices c and d may be true statements, but do they relate to the question at hand?
For question 20:	This is about a test cross.

CONGRATULATIONS . . . YOU HAVE COMPLETED CHAPTER 7 !!!

CHAPTER 8

ADVANCES IN GENETICS

CONNECTIONS

It has been said in the past that, since you are stuck with the genes you received from your parents, there is nothing you can do about the way you are. And, except for science fiction, behavior modification, cosmetics, and surgery it was mostly true. At this point in the course you understand the principles of metabolism and genetics that underlie these restrictions.

Chapter Eight reminds you that we are now witnessing an information explosion in biology. Technological advances are providing genetic engineers with the tools to change and control the genetics of many organisms. Look for ways that scientists have made use of natural genetic mechanisms to produce new results. As you read, consider how much your opinion of this topic is based on facts and how much is based on emotion.

OBJECTIVES FOR THIS CHAPTER

When you have finished Chapter 8, you should be able to:

1. Briefly outline the steps in forming recombinant DNA.
2. Describe in detail the role of restriction enzymes in genetic engineering.
3. Explain how vectors and plasmids are used to clone a gene.
4. Explain why bacteria are used as vectors.
5. Discuss the advantages and possible problems associated with genetic engineering.
6. List ways in which genetic engineering techniques are used in agriculture and medicine.
7. Describe the advances made in gene replacement therapy.
8. Explain why gene replacement therapy may be more advantageous to humans than relying only on recombinant DNA production in bacteria.

CHAPTER 8 OUTLINE

I. Genetic Engineering: An Unfolding Story
   A. The Players
   B. Finding and Isolating the Gene
   C. How DNA Is Chopped Up
   D. Getting Genes from One Place to Another
   E. Splicing and Cloning the Genes
   F. Sorting Through the Recombinants
II. Is It Tampering with Genes or Genetic Engineering?
III. How Great a Promise?
   A. Advances in Agriculture
   B. Diagnosis and Decision in Humans
   C. Gene Replacement Therapy

CONCEPTS IN BRIEF

Genetic Engineering: An Unfolding Story

An organism uses whatever genetic material it has, and produces proteins accordingly. If a new piece of DNA is inserted into the genetics of the cell the new genetics will be read and utilized as if it were there all along. Since it is relatively easy to put new genes into the plasmids of bacterial cells, scientists use them for much of the genetic engineering work that is done.

Creating recombinant DNA (the original host DNA with added DNA) in a cell requires many steps. Restriction enzymes are used to break apart strands of DNA where specific sequences of nucleotides are found. The broken (sticky) end is now a good site for attaching the desired (foreign) segment of DNA as long as it has the same nucleotide sequence on its broken end as well. The foreign DNA may come from another cell (possibly even from a different type of organism) or it may have been synthesized in a lab. The two types of DNA are mixed together with the enzyme ligase, and will attach their sticky ends to one another.

Large batches of bacteria are used for these experiments, and only a fraction of the cells form recombinant DNA. These cells are sorted out from the rest and are grown in pure cultures called clones. They should now form the proteins prescribed by the inserted DNA.

Is It Tampering with Genes or Genetic Engineering?

The prospect of genetic engineering has caused a great deal of speculation about the applications of such a technology. Scientists and government agencies have planned their research with careful controls in order to avoid releasing potentially harmful organisms into the environment. The opportunity for financial rewards from medical and agricultural advances has led to a rapid increase in genetic engineering research.

How Great a Promise?

Scientists are working to improve crop and livestock production by increasing yields and decreasing disease and pest problems. Applications to humans have centered around using bacteria to manufacture chemicals which are used as treatments for conditions caused by genetic defects. In addition, tests have been developed which provide information for family planning. Much current research is aimed at being able to actually *replace* defective genes in people so that they no longer have defective genes.

KEY TERMS

Be sure to make and use flash cards for all of these terms. Suggestions are found in "To The Student" at the beginning of the Study Guide. Page numbers refer to the text.

antibiotics	212	host DNA	210
clone	210	plasmid	212
conjugation	214	recombinant DNA	210
E. coli	211	restriction enzyme	210
gene replacement therapy	222	source DNA (foreign DNA)	210
genetic engineering	216	vector	210

EXERCISES

Check your understanding of Chapter 8 by completing the following exercises. Answers begin on p. 207.

1. List five possible improvements in crops that could be achieved through genetic engineering.
   a.
   b.
   c.
   d.
   e.

True or False Questions

Mark each question either T (True) or F (False).

____2. Under normal conditions, bacterial cells contain plasmids, but human cells do not.

____3. Recombinant DNA may contain genetic material from two different species.

____4. Bacterial cells are commonly used as vectors for recombinant DNA.

____5. Restriction enzymes occur naturally in bacteria.

Matching Questions

Each question may have more than one answer; write all of the answers that apply. Answers may be used more than once.

____6. small, circular, bacterial DNA molecule	a. plasmid
____7. a population of identical organisms	b. vector
____8. composed of DNA from two different sources	c. host DNA
____9. *E. coli* is a common example	d. recombinant DNA
____10. transferred during bacterial conjugation	e. clone

Multiple-Choice Questions

Circle the choice that is the best answer for each question.

11. If the sticky end of a plasmid DNA had the nucleotide sequence GCT, what nucleotide sequence would the foreign DNA need to have on its sticky end?
    a. GCT
    b. TCG
    c. CGA
    d. AGC
    e. CGU

12. Scientists can separate the bacteria with altered plasmids from those which do not have them by exposing the bacteria to
    a. antibiotics.
    b. calcium chloride.
    c. recombinant DNA.
    d. vectors.
    e. glycophosphate.

13. During the process in which bacteria recombine their genes, the _____ forms a _____ tube.
    a. F^+ cell, female
    b. F^- cell, male
    c. F^+ cell, conjugation
    d. F^- cell, conjugation

14. Instead of using livestock as a source of insulin, people can now get _____ insulin that is produced by bacteria.
    a. bacterial
    b. cow
    c. human
    d. plant
    e. all of the above

15. If you had gene replacement therapy done on yourself, and you wanted the results to affect your offspring, it would be necessary to be sure to alter the genes in
    a. your entire body.
    b. your reproductive organs.
    c. the plasmids of the affected organ.
    d. the recombinant DNA of the *E. coli* in your offspring.

Fill in the Blanks

Complete each statement by writing the correct word or words in every blank.

16. Host DNA is cut by a __________ enzyme at a specific site. It can then be joined to __________ DNA by the enzyme __________. The plasmid can now be said to contain __________ DNA.

17. A__________ can be grown from a single bacteria that contains a plasmid with recombinant DNA.

Questions For Further Thought

Write your answer for each question in the space provided.

18. Assuming that the technological problems had been solved, discuss the advantages of using gene replacement therapy in humans over relying on treating medical problems with the chemicals which can be produced from recombinant DNA in bacteria.

19. Explain why it was important to do genetic engineering experiments with bacteria that could not survive in the environment.

SUMMARY

Study the seven items listed in the SUMMARY on p. 223 of the text. It should be clear to you that scientists are now able to manufacture many kinds of DNA segments and insert them into cells where they will function as part of the cell's normal metabolism. This new DNA will, of course, change the nature of the cell. You can see that the potential changes which could result from genetic engineering have many ramifications. It is important that the applications of this technology are used wisely.

HINTS

Read these suggestions before you check your answers to the EXERCISES.

For question 2: Don't forget that *you* have chromosomes.

For question 10: Could recombinant DNA have been formed before conjugation?

For question 11: Remember your DNA base pairing.

For question 14: Does it depend on the kind of cell it comes from, or is it the kind of DNA that is important?

For question 15: Where do sperm and eggs come from in your body?

CONGRATULATIONS . . . YOU HAVE COMPLETED CHAPTER 8 !!!

Part Three The Diversity of Changing Life

CHAPTER 9

THE FIRST LIFE

CONNECTIONS

It has probably been obvious to you for a long time (at least since the beginning of the course) that there are so many different kinds of living things on Earth that you are never going to see them all. This chapter examines the current scientific theories of how life of all types came to be on our planet. Once again "life" is not defined, it is simply described.

This is the first of five chapters (Part Three of the text) in which you will study the characteristics of the major groups of organisms around us. Watch for the ways in which all life forms are similar to, yet different from each other. Also keep in mind that throughout time, the characteristics of living things have been changing.

OBJECTIVES FOR THIS CHAPTER

When you have finished Chapter 9, you should be able to:

1. Compare and contrast the views of mechanists and vitalists on the nature of life.
2. Describe how the hypothetical "Big Bang" accounts for the formation of the universe.
3. Compare the environmental conditions of early earth with the earth today.
4. Describe the experiments of Stanley Miller and explain how his results affected current scientific thinking on how life might have arisen.
5. Describe the experiments of Sidney Fox and explain how his results affected current scientific thinking on how life might have arisen.

6. Trace the steps by which microspheres, liposomes, or coacervate droplets may have evolved into living cells.
7. Describe the factors which affected the evolution of early autotrophs and heterotrophs.
8. Describe how the presence of oxygen changed the world and its early life forms.

CHAPTER 9 OUTLINE

I. Life from nonlife
 A. Mechanism vs. Vitalism
 B. Life's Many Meanings
II. The Expanding Universe
III. Our Early Earth
IV. Hypotheses Concerning the Origin of Life
V. The Formation of the Molecules of Life
VI. Protobionts: The Ancestral Droplets
 A. Reproducing Droplets
 B. Successful Droplets
VII. Heterotrophy and Autotrophy
VIII. Oxygen: A Bane and a Blessing

CONCEPTS IN BRIEF

Life from Nonlife

Explanations of life's origins on earth have been given in many forms. Scientists have commonly fallen into either the mechanist or vitalist camp. Since vitalism isn't testable by scientific experimentation (although it may seem easier and more fun than mechanism), this chapter looks at mechanistic explanations of life. This viewpoint says that you can learn all about a living thing if you understand its chemistry and physics. Consequently you should be prepared to keep using the chemistry you have already learned.

The Expanding Univers

All the matter of our universe may once have taken up less room than you do, before it exploded in the "Big Bang" over thirteen million years ago. Since then, the universe has been expanding and the structures that we see today have been formed from what was in that original small lump.

Our Early Earth

The atmosphere of the early earth did not contain any free oxygen (O_2). Therefore, most modern forms of life could not have survived in the conditions of the early earth. If there *had* been free oxygen available, however, the chemical reactions required to develop life could not have occurred.

Hypotheses Concerning the Origin of Life

Life forms (at least simple ones which left fossils) have been on the earth for at least 3.5 billion years. It is thought that various energy sources were available and were able to support reactions between molecules that had become concentrated in the developing oceans. These reactions may have led to the kinds of compounds found in modern life forms.

The Formation of the Molecules of Life

Stanley L. Miller began his research in this area in 1953. He experimented with the gasses that were thought to have been in the early atmosphere and discovered that they could spontaneously form amino acids. Since then his techniques have been used (and modified by various researchers) to form many biologically important compounds.

Sidney Fox has shown that complex molecules (proteinoids) could be formed in primitive conditions from the kinds of compounds that come out of a Miller-type apparatus.

Protobionts: The Ancestral Droplets

Life probably developed from a type of droplet structure. By having an external chemical boundary that separated the interior from the environment, the structure could: (1) regulate what passed into and out of itself, and (2) hold concentrations of chemicals inside itself where they had some degree of protection from the environment. Such a structure could also grow and divide, or break apart and form additional droplets, thereby reproducing.

Suggested structures which could have done these things include A. I. Oparin's coacervate droplets, Sidney Fox's microspheres, and liposomes. The successful structures, or protobionts, undoubtedly reproduced in organized ways that resulted in "daughters" that were similar to the original protobiont structure. At some point in time these protobionts had acquired the characteristics that we attribute to cells.

Heterotrophy and Autotrophy

The earliest cells were heterotrophs, that is they took in chemicals from their environment as food. Some cells would have had the chemistry that allowed them to manufacture additional compounds from those which they could take in. As their chemical pathways became more refined, cells which could make their own food (autotrophs) were successful as well. These cells would have made a good food source for other cells, and thus, feeding relationships developed.

Oxygen: A Bane and a Blessing

You may recall from Chapter Five that oxygen is released during photosynthesis, but don't forget that free oxygen was not a part of the early environment yet. Once free oxygen accumulated in the atmosphere, ozone formed and acted (as it still does today) to filter harmful ultraviolet rays from the sun. This made it safer for organisms to occupy the land environments of the earth.

KEY TERMS

Be sure to make and use flash cards for all of these terms. Suggestions are found in "To The Student" at the beginning of the Study Guide. Page numbers refer to the text.

autotroph	241	mechanism	229
carnivore	242	microsphere	237
coacervate droplet	237	omnivore	242
herbivore	242	proteinoid	236
heterotroph	241	protobiont	237
liposome	237	vitalism	229

EXERCISES

Check your understanding of Chapter 9 by completing the following exercises. Answers begin on p. 207.

1. Arrange the following events in sequence, from earliest to most recent.
   a. free oxygen is present in the atmosphere
   b. Miller produces amino acids
   c. autotrophs develop
   d. small biological molecules form polymers
   e. heterotrophs appear
   f. formation of the earth

_____ _____ _____ _____ _____ _____

Earliest   Most Recent

True or False Questions

Mark each question either T (True) or F (False).

____2. If you make coacervates in a lab, they can be considered living cells.

____3. The outer boundary of a coacervate acts somewhat like a plasma membrane acts in a cell.

____4. Our research into the origin of cells has given us a clear definition of life.

____5. The first living cells must have been anaerobic.

____6. The Big Bang explanation of the origin of the universe says that all the matter of the universe was once all together in a small lump.

Matching Questions

Each question may have more than one answer; write all of the answers that apply. Answers may be used more than once.

____7. acquire their food from the environment around themselves

____8. eat only heterotrophs

____9. eat heterotrophs and autotrophs

____10. are living on the earth today

a. autotroph
b. heterotroph
c. omnivore
d. herbivore
e. carnivore
f. none of the above

____11. was present in the early atmosphere

____12. has been produced in a Miller-style apparatus

____13. reacted with other molecules of the same kind to produce ozone

____14. was probably used in energy reactions by early cells free oxygen

a. free oxygen
b. amino acid
c. hydrogen
d. ATP
e. purine
f. none of the above

Multiple-Choice Questions

Circle the choice that is the best answer for each question.

15. Most scientists today consider themselves to be
    a. protobionts.
    b. coacervatists.
    c. mechanists.
    d. vitalists.

16. Sidney Fox showed that when__________are mixed with water in a laboratory they assemble into__________.
    a. protobionts, cells
    b. proteinoids, cells
    c. protobionts, microspheres
    d. proteinoids, microspheres

17. The oldest fossil remains which have been found of photosynthetic organisms show that they lived at least _____ years ago.
    a. 20,000
    b. 1.5 million
    c. 3.5 billion
    d. 13 billion

18. Before the appearance of protobionts, there were _____ living things on the earth.
    a. no
    b. very few
    c. heterotrophic but no autotrophic
    d. only aquatic

19. The first cells were probably
    a. autotrophs.
    b. heterotrophs.
    c. mechanists.
    d. vitalists.

20. The development of a layer of ozone made it
    a. more difficult to be an autotroph.
    b. easier to survive on the land.
    c. possible for microspheres to become coacervates.
    d. more difficult for carnivores to manufacture their own food.

Fill in the Blanks

Complete each statement by writing the correct word or words in every blank.

21. A gas that was not present in the atmosphere of the primitive earth was _________.

22. Stanley _________ put gasses thought to have been found in the early earth's _________

in a glassware apparatus in 1953. He used __________ as an energy source, and over a period of

several days __________ were formed.

23. Vitalists say that living things are different from nonliving things because they have a __________.

24. Droplet structures have been proposed as being important in the development of cells. Examples

of these include __________, __________, and __________.

Questions For Further Thought

Write your answer for each question in the space provided.

25. Explain why the development of life on the earth required a lack of free oxygen in the beginning, yet also depended on the appearance of free oxygen later.

26. Discuss why it would be beneficial for protobionts to reproduce "daughters" that were chemically *similar* to the original.

SUMMARY

Study the ten items listed in the SUMMARY on p. 247 of the text. You have seen that life began through a series of changes involving chemical systems in the waters of the earth. As accumulations of these chemicals became more complex they began to show the characteristics that we attribute to living things. Those pre-cells continued to change, and from them developed the ancestors of all the living things on the earth today.

HINTS

Read these suggestions before you check your answers to the EXERCISES.

For question 2: This refers to doing this today, not on the ancient earth.

For question 4:	This asks about a definition, not a description.
For question 5:	Was there oxygen available then?
For question 7:	Remember, there are several kinds of heterotrophs.
For question 12:	This has three correct answers.
For question 16:	Remember when Fox did his work and when protobionts were around.
For question 18:	This is asking about *living* things.
For question 20:	Think about what ozone filters out.
For question 21:	What you have learned about mitosis will apply to this question as well.

CONGRATULATIONS . .. YOU HAVE COMPLETED CHAPTER 9 !!!

CHAPTER 10

THE PROCESSES OF EVOLUTION

CONNECTIONS

Now that you have finished the previous chapter you are aware that living things have been changing since they came into existence. Think back to Chapter One and Darwin's concept of natural selection. Before the 1940s, natural selection was the primary explanation for the mechanism of the process of evolution. By then, however the concepts of genetics (as well as some ecology) had been integrated with natural selection and a more complete theory of evolution (the modern synthesis) developed.

This chapter presents a more thorough look at evolution than you have already seen in the course. Watch for the factors which are responsible for changes in populations (*not* in individuals) over time. In the chapters to come you will be studying the results of these processes of change as you examine the diverse types of living things found on the earth today.

OBJECTIVES FOR THIS CHAPTER

When you have finished Chapter 10, you should be able to:

1. Explain the differences between Darwin's concepts of evolution and the modern synthesis.
2. State how the modern synthesis accounts for evolution within populations.
3. Describe how variations in populations are produced and maintained by the environment.
4. Explain how differences in populations are produced by stabilizing, directional, and disruptive selection.
5. Describe two mechanisms that bring about genetic drift in small populations.
6. Discuss why the effects of genetic drift are so apparent in *small* populations.
7. Explain some difficulties encountered in defining the term "species"
8. List several reproductive isolating mechanisms and their effects.
9. Compare and contrast allopatric and sympatric speciation.
10. Recount several hypotheses for the causes of species extinction during the Cretaceous period.

CHAPTER 10 OUTLINE

I. Variation in Populations
   A. Variation by Single-Gene and Polygenic Inheritance
   B. How the Environment Maintains Variation
   C. Factors Involved in Evolution
   D. How Natural Selection Can Influence Variation
      1. Modes of Natural Selection
   E. Small Populations and Genetic Drift
      1. Bottleneck Effect
      2. Founder Effect
II. The Question of Species
III. Speciation
   A. Allopatric Speciation
   B. Sympatric Speciation
   C. Divergent and Convergent Evolution
IV. Small Steps or Great Leaps?
V. Death Stars and Dinosaurs: The Great Extinctions

CONCEPTS IN BRIEF

Variation in Populations

You already know that the individual members of populations are not all alike; variation in populations is normal and expected. Several factors can affect the appearance of individuals, but only mutation can introduce new alleles. Higher levels of variation within a population make it more likely that at least some members of a population will survive as the environment changes.

Many factors influence the variation that is found in any population. These factors include:

- how many genes control the expression of a trait
- whether the environment is stable or unstable
- the frequencies of alleles in the population
- mutation
- movement of individuals into and out of the population
- the reproductive success of individuals with different genotypes
- the type of natural selection acting on the population
- the size of the population
- random events in nature
- natural disasters.

Notice that the items in this list center around: (1) factors which violate the requirements of the Hardy-Weinberg law and (2) the interaction of the population and the environment.

The Question of Species

Reproductive isolation is the key element in Ernst Mayr's definition of species. The five types of reproductive isolation describe which factors may act to prevent the members of two populations from successfully reproducing with one another.

Speciation

New species are formed from existing species. If the original group has been physically separated into two groups, the process is called allopatric speciation. This seems to be the most common way in which new species develop. However, if all the members involved stay together in the same place during the process, the term sympatric speciation is applied. Plants exhibit this method more often than animals.

Different species which live in ecologically similar conditions may look very much alike even though they are not closely related to one another.

Small Steps or Great Leaps?

A recent interpretation of the fossil record by Niles Eldridge and Stephen Jay Gould has led them to suggest that, throughout history, species may have remained more or less the same for long periods of time. When changes did happen, they occurred more suddenly than had been thought.

Death Stars and Dinosaurs: The Great Extinctions

Around sixty five million years ago there were mass extinctions of many groups of organisms on Earth. Suggested causes include a lowering of the earth's temperature, a collision with an asteroid which raised enough dust to block the sun and inhibit photosynthesis, and collisions with asteroids because of the gravitational influence of a sister "death star." You can probably rest assured that none of these will be repeated before the end of this course!

KEY TERMS

Be sure to make and use flash cards for all of these terms. Suggestions are found in "To The Student" at the beginning of the Study Guide. Page numbers refer to the text.

allopatric speciation	266	gene pool	252
Alvarez hypothesis	275	genetic drift	260
bimodal curve	259	hybridization	268
bottleneck effect	262	polygenic inheritance	254
cline	256	polymorphic	260
convergent evolution	269	punctuated equilibrium	270
dimorphic	260	reproductive isolation	265
directional selection	258	sexual dimorphism	261
disruptive selection	259	speciation	266
divergent evolution	269	species	265
evolution	251	stabilizing selection	258
founder effect	262	sympatric speciation	268

EXERCISES

Check your understanding of Chapter 10 by completing the following exercises. Answers begin on p. 207.

1. Assume that you are a botanist and you have been studying a population of purple flowers. When you began the study ten years ago there were a few light flowers, a few dark ones, and the majority were a medium purple color. Over time there has been a gradual change. Now the majority are a light purple color, there are a few medium ones, and there are almost no dark ones.
Draw two graphs which show the distribution of colors at the beginning and end of your study.

Early Population   Current Population

True or False Questions

Mark each question either T (True) or F (False).

_____2. Meiosis and crossing over are methods of adding new alleles to a population.

_____3. The founder effect is more likely to be seen in small populations than in large populations.

_____4. The fossil evidence clearly shows that evolutionary changes are gradual and require long spans of time.

_____5. Members of the same species, which live in the same area, would normally produce fertile offspring.

_____6. A change on the gene pool would be considered to be evolution.

_____7. Reproductive isolation requires that two populations be separated by a physical barrier.

Matching Questions

Each question may have more than one answer; write all of the answers that apply. Answers may be used more than once.

_____8. is able to evolve

_____9. has a particular set of alleles

_____10. must be able to survive in its environment

a. an individual organism
b. a population

____11. occurs when the environment does not change over long periods of time	a. directional selection b. disruptive selection c. stabilizing selection
____12. the result of one characteristic having a higher rate of survival	
____13. may produce a population with two different successful types	
____14. illustrated by Question #1 above	

Multiple-Choice Questions

Circle the choice that is the best answer for each question.

15. If strong selection is acting on a population, the result will be _____ in the variation within the population.
   a. a decrease
   b. an increase
   c. no apparent change

16. When the effects of the environment result in two unrelated organisms having a similar appearance, you would say that _____ has occurred.
   a. stabilizing selection
   b. sympatric speciation
   c. convergent evolution
   d. punctuated equilibrium

17. The fact that different colors of the English peppered moth were more common before and after industrialization is a good example of
   a. directional selection.
   b. founder effect.
   c. a bimodal curve.
   d. sympatric speciation.

18. It is advantageous for _____ populations to show variation.
   a. small
   b. large
   c. isolated
   d. all

19. The fittest organisms are those which
   a. are the strongest.
   b. do not have a gene pool.
   c. produce the most live offspring.
   d. have the highest rates of convergent evolution.

Fill in the Blanks

Complete each statement by writing the correct word or words in every blank.

20. If a population is evolving, it must be violating the conditions of the __________ law. As a result, the frequencies of its __________ would be changing over time.

21. The modern synthesis is a combination of Darwin's ideas and the concepts of __________ __________.

22. Genetic drift affects __________ populations. Two examples of this are the __________ effect due to natural disasters, and the __________ effect when a small portion of the population is separated from the main body of the population.

23. The Alvarez hypothesis states that late in the __________ period, the __________ collided with a large __________. As a result, large clouds of __________ blocked the sunlight and plant __________ was disrupted. This affected many food chains.

Questions For Further Thought

Write your answer for each question in the space provided.

24 Even though the environment may cause variations that can be seen in a population, this is not an example of evolution. Explain why.

25. Give some reasons why sympatric speciation is not as common in animals as it is in plants.

SUMMARY

Study the twenty-one items listed in the SUMMARY on p. 277 of the text. Notice that the emphasis in the chapter has been on populations (rather than individuals) and their interaction with the environment. Although previous chapters have dealt mostly with individuals, it is important to keep in mind that evolution occurs when the allele frequencies of *populations* change; evolution does not occur to individuals. Populations will only stop evolving if the conditions of the Hardy-Weinberg law are met.

Long term survival of a species requires plenty of successful reproduction and enough variation to allow adaptation to environmental changes. However, if environmental changes happen too rapidly, species may become extinct. At the moment, humans are the major cause of extinctions world wide.

HINTS

Read these suggestions before you check your answers to the EXERCISES.

For question 2: The key word here is "new."

For question 7: There is more than one kind of reproductive isolation.

For question 15: This is about strong selection, not weak selection.

For question 16: These organisms are *not* related.

For question 17: Remember, both types of moth were there all along.

For question 19: This is asking about fitness (being the fittest) in *evolutionary* terms.

For question 24: Don't start giving the explanations that Lamarck would have given.

CONGRATULATIONS . . . YOU HAVE COMPLETED CHAPTER 10 !!!

CHAPTER 11

DIVERSITY I: THE MONERA, PROTISTA, AND FUNGI

CONNECTIONS

Up to this point you have learned several important principles (and numerous details) of biology which you can apply to all organisms. If you have been waiting to learn about the structures and functions of specific kinds of living things, your wait is finally over!

Even though biologists cannot always agree on the appropriate criteria, the classification of living things is an important area of study. This chapter introduces one system that is commonly used to identify all life forms on Earth. In addition, the members of the kingdoms Monera, Protist, and Fungi are examined in some detail. Look for the common features which can be found in all of these groups, as well as for the unique characteristics which are used to distinguish them from one another.

OBJECTIVES FOR THIS CHAPTER

When you have finished Chapter 11, you should be able to:

1. Name the five kingdoms into which living things are grouped.
2. List the characteristics that are used to separate organisms into five kingdoms.
3. List the different types of monerans and describe their diverse habitats.
4. Discuss the features of the cyanobacteria that make them unique monerans.
5. Give examples of ways in which humans are affected by monerans.
6. Describe the features of the animallike, plantlike, and funguslike protists.
7. Explain the relationships of the protists to the other eukaryotic kingdoms.
8. Give examples of ways in which humans are affected by protists.
9. Describe representative examples of the four divisions of fungi.
10. List ways in which fungi are beneficial and detrimental to humans.
11. Relate examples of symbiotic relationships shown by fungi.

CHAPTER 11 OUTLINE

I. Five Kingdoms and Splitters and Lumpers
II. The Kingdom Monera and a Continuing Simplicity
    A. Division Archaebacteria and Extreme Habitats
    B. Eubacteria and a Sea of Life
III. The Kingdom Protista and Accelerating Evolution
IV. The Animal-Like Protists
    A. Phylum Ciliophora and Tiny Predators
    B. Phylum Sarcodina and a Flexible Simplicity
    C. Phylum Sporozoa and a Parasitic Existence
V. The Plantlike Protists
    A. Phylum Euglenophyta and an Evolutionary Puzzle
    B. Phylum Chrysophyta and Glassy Corpses
    C. Phylum Pyrrophyta and Sparkling and Dreaded Waters
VI. The Fungus-Like Protists
VII. The Kingdom Fungi and Varied Lifestyles
    A. Division Oomycota and Irish Americans
    B. Division Zygomycota and Preserving Our Bread
    C. Division Ascomycota and Remarkable Partnerships
    D. Division Basidiomycota and the Destroying Angel
    E. Division Deuteromycota--The Fungi Imperfect
    F. Mycorrhizae and Associating with Plants

CONCEPTS IN BRIEF

Five Kingdoms and Splitters and Lumpers

When living things are named they are grouped with related organisms, although biologists cannot always agree on how *closely* some things are related. Most scientists have agreed to use a five kingdom system of classification. That means that each living thing belongs in one of the five kingdoms. Then it is placed into one of the possible phyla or divisions of that kingdom, and in one of the possible classes, etc. The genus and species names are usually used when discussing a particular organism.

The Kingdom Monera and a Continuing Simplicity

Prokaryotic organisms are all single-celled and have been classified into the Kingdom Monera. They do not contain membranous organelles, such as ribosomes or a nucleus. The archaebacteria often live in extreme habitats where few other organisms can survive. Eubacteria are more common and are responsible for food spoilage, decomposition, nitrogen fixation, numerous diseases, and the production of many antibiotics. Bacteria commonly form resistant endospores as a way of surviving unfavorable conditions such as acidic environments, high temperatures, and dryness.

The Kingdom Protista and Accelerating Evolution

Members of the Kingdom Protista are eukaryotic, therefore their cells are like those that you studied in Chapter Four. Many of them are put into this kingdom because it is clear that they don't belong in any of the other four kingdoms, rather than because of any particular relationship. Their ancestors were probably the protists.

The Animallike Protists

Before the five kingdom system was adopted, all living things were put into two kingdoms: animals or plants. All the protists that were in the animal kingdom are now informally grouped together as the animallike protists. Ciliophora all use cilia for moving about and capturing food. Paramecia are common examples. Sarcodina are unique because of their amoeboid motion in which they seem to flow through the environment. Like the Ciliophora, they regulate water by the means discussed in Chapter Four. The sporozoans are parasitic, relying on other organisms for their survival. Malaria is caused by the sporozoan *Plasmodium.*

The Plantlike Protists

As you probably guessed, the plantlike protists were once part of the plant kingdom and are photosynthetic. Members of the Euglenophyta use one or more long flagella to move through the environment. Their ability to photosynthesize places the *Euglena* here, even though they also have several animal-like characteristics. The cyst stage of protists acts in the same protective way as the bacterial endospore.

Diatoms (Phylum Chrysophyta) are important because they are a source of food for many other organisms. Their glass-like coverings accumulate at the bottom of the sea, and are found in deposits of diatomaceous earth. Both toxic red tides and water that glows at night are the result of members of the Pyrrophyta, commonly called dinoflagellates.

The Funguslike Protists

The slime molds were, of course, once in the Kingdom Fungi but are now considered to be protists. Both the cellular and acellular types have very unusual life cycles in which they show characteristics that resemble those of animals, fungi, and protists.

The Kingdom Fungi and Varied Lifestyles

Fungi have been separated from the plants because of: their hyphae, their cell walls that contain chitin, their lack of chlorophyll, and their parasitic and saprophytic method of digesting food outside of their bodies and then absorbing the results. Most fungi (except for the Deuteromycota) can reproduce sexually, but asexual reproduction is more common.

Haploid spores are a common feature of the reproductive cycles, and they are often encased in protective structures like the cysts and endospores you have already read about. Fungi are used as food, are responsible for decomposition, and cause many important diseases. Lichens and mycorrhizae are the result of a very close and beneficial association of specific fungi with algae and plants.

KEY TERMS

Be sure to make and use flash cards for all of these terms. Suggestions are found in "To The Student" at the beginning of the Study Guide. Page numbers refer to the text.

archaebacteria	285	basidiocarp	309
ascocarp	306	basidiomycetes	307
ascomycetes	302	basidiospores	310
ascospores	306	basidium	309
ascus	306	Chrysophyta	297
bacillus	288	cilia	292

Term	Page
Ciliophora	292
coccus	288
conidia	303
conidiophore	306
conjugation	294
contractile vacuole	294
cyst	297
cytopyge	294
cytostome	293
deuteromycetes	310
dinoflagellate	298
ectoplasm	295
endoplasm	295
endospore	288
eubacteria	286
Euglenophyta	295
eyespot	297
flagellum	295
food vacuole	293
fungi	300
gullet	297
hypha	301
macronucleus	294
micronucleus	294
moneran	285
mycelium	301
mycorrhizae	311
oomycetes	301
oral groove	293
Paramecium	292
pellicle	292
Protista	292
protozoan	292
pseudopod	295
Pyrrophyta	298
Sarcodina	294
spirillum	288
Sporozoa	295
taxonomist	284
trichocysts	294
zygomycetes	303
zygospore	303

EXERCISES

Check your understanding of Chapter 11 by completing the following exercises. Answers begin on p. 207.

1. Arrange the seven levels of taxonomy in sequence from most inclusive to least inclusive.

_____ _____ _____ _____ _____ _____ _____

Most Inclusive ... Least Inclusive

2. While you were out tromping around you discovered an organism that is multicellular, has chitin in its cell walls, and produces ATP with its mitochondria.
   a. Does this fit into one of the current kingdoms? _____

   b. If so, which one? __________

True or False Questions

Mark each question either T (True) or F (False).

_____3. The current system of taxonomy places non-cellular organisms into the Kingdom Monera.

_____4. The Basidiomycota produce haploid spores.

_____5. Most biologists classify all organisms with a five kingdom system.

____6. Fungal spores are usually delicate and are easily killed.

____7. When fungi reproduce sexually, the hyphae of different mating types fuse.

Matching Questions

Each question may have more than one answer; write all of the answers that apply. Answers may be used more than once.

____8. may contain chloroplasts

____9. at least some have cell walls

____10. have a cell nucleus

____11. have prokaryotic cells

____12. some live in hot springs

a. Kingdom Monera
b. Kingdom Protista
c. Kingdom Fungi
d. all of the above
e. none of the above

____13. shows amoeboid movement

____14. causes red tides

____15. can photosynthesize

____16. are diatoms

a. Ciliophora
b. Sarcodina
c. Sporozoa
d. Euglenophyta
e. Chrysophyta
f. Pyrrophyta
g. all of the above

Multiple-Choice Questions

Circle the choice that is the best answer for each question.

17. The monerans may have been the ancestors to the
    a. eukaryotes.
    b. prokaryotes.
    c. viruses.
    d. all of the above.

18. The deuteromycetes are known for
    a. having specialized flagella.
    b. their apparent lack of sexual reproduction.
    c. the fact that they often form symbiotic relationships with algae.
    d. all of the above.

19. A paramecium releases excess water through
    a. a contractile vacuole.
    b. amoeboid motion.
    c. its conidiophores.
    d. specialized mycorrhizae.

20. Lichens are
    a. composed of a fungus and an alga.
    b. used as food by some organisms.
    c. generally slow growers.
    d. common in the arctic.
    e. all of the above.

Fill in the Blanks

Complete each statement by writing the correct word or words in every blank.

21. Fungi grow by producing long __________ that join together (often under ground or in a host) to form a __________. In the Basidiomycota, a __________ __________ such as a mushroom may appear above the ground.

22. Some bacteria, such as the __________ group, produce __________ which kill other bacteria.

23. All prokaryotic organisms are placed into the Kingdom __________.

24. All of the members of the protist phylum __________ are parasitic.

25. Mitochondria probably developed in the Kingdom __________. This was an important step in the efficient use of __________ by cells.

Questions For Further Thought

Write your answer for each question in the space provided.

26. The Kingdom Protist is sometimes used as a category for organisms that do not fit into another kingdom. Explain how *Euglena* illustrates this.

SUMMARY

Study the twenty-four items listed in the SUMMARY on p. 311 of the text. As you completed this chapter, you probably realized that you have never seen most of the organisms mentioned in this chapter, although you were familiar with some of their activities and effects. The monerans and protists are usually too small to notice, and most of the body of a fungus is hidden.

A great deal of work remains to be done in order to understand how the members of these kingdoms are related to one another. Regardless of our lack of understanding, their photosynthesis, decomposition,

decomposition, nitrogen fixation, and abilities to cause and cure diseases make them very important organisms in the environment, as you have seen.

HINTS

Read these suggestions before you check your answers to the EXERCISES.

For question 3:	Only cellular organisms are found in the five kingdoms.
For question 8:	Don't forget the differences between prokaryotic and eukaryotic cells.
For question 10:	See the hint for Question #8.
For question 17:	It wants to know what group did the monerans give rise to.
For question 25:	Mitochondria are organelles and are not found in prokaryotic cells.

CONGRATULATIONS . . . YOU HAVE COMPLETED CHAPTER 11 !!!

CHAPTER 12

DIVERSITY II: THE PLANTS

CONNECTIONS

When you think of plants you probably associate them with photosynthesis and the release of oxygen, along with a variety of other ways in which you find them pleasant and useful. This chapter continues to examine the ways in which plants are different from the other kingdoms, and what has made them such successful organisms.

The emphasis here is on the particular characteristics of the divisions of plants. The details of plant structure and function will be presented in Chapter Fourteen. Look for the evolutionary significance of various traits as you read. Think about how certain characteristics of each division are utilized to enhance the survival of that group in its environment.

OBJECTIVES FOR THIS CHAPTER

When you have finished Chapter 12, you should be able to:

1. Describe the alternation of the gametophyte and sporophyte generations in plants.
2. Explain why the alternation of generations cycle may have evolved.
3. Give the characteristics of the three algal divisions.
4. Tell which algal division is the most likely ancestor of land plants, and present the evidence.
5. List three types of bryophytes and explain how they differ from vascular land plants.
6. List the divisions of vascular land plants and describe their special adaptations to terrestrial life.
7. Describe the evolutionary advances that are characteristic of ferns.
8. List the divisions of gymnosperms and tell why they are grouped together.
9. Differentiate between gymnosperms and angiosperms.
10. List the evolutionary trends in flower development.

CHAPTER 12 OUTLINE

I. Alternating Generations
   A. The Evolution of Alternating Generations

II. Aquatic Plants
   A. Division Rhodophyta and Many Colors
   B. Division Phaeophyta and Undersea Forests
   C. Division Chlorophyta: Ancestors of Land Plants
III. Nonvascular Land Plants
   A. Division of Bryophyta and the Invasion of the Land
IV. Vascular Plants
   A. Division Pterophyta and Leaves and Roots
   B. Gymnosperms and Naked Seeds
   C. Angiosperms, Covered Seeds, and Flowers

CONCEPTS IN BRIEF

Alternating Generations

If you sat and watched the phases in the life cycle of a plant you would see it demonstrate alternation of generations. A sporophyte would be followed by a gametophyte which would be followed by a sporophyte, etc. In most species the sporophyte generation looks different from the gametophyte generation. You may never have noticed that these stages exist, since one of the stages is often inconspicuous or short lived.

Remember what you have learned about mitosis (followed by cytokinesis) and meiosis. Keep the following facts about plants in mind:

- plants in the sporophyte generation are diploid
- plants in the gametophyte generation are haploid
- spores are haploid
- gametes are haploid
- the sporophyte generation makes spores
- the gametophyte generation makes gametes.

Alternation of generations may be beneficial to plants by: (1) providing a mechanism for mixing alleles from different parents (remember meiosis and independent assortment?), and (2) affording a chance to respond to environmental stress.

Aquatic Plants

The red, brown, and green algae are referred to as the aquatic plants, even though some of them live for long or short periods on land. All of them lack special structures for transporting food and water inside their bodies. Because of this, their body parts cannot be called stems, roots, or leaves.

Some of the Rhodophyta are able to survive in very deep water, since their phycoerythrin pigment captures blue light and transfers the energy to chlorophyll for photosynthesis. Members of the Phaeophyta can be very large. Their air-filled bladders are useful for buoyancy and support.

Modern land plants are probably descended from members of the phylum Chlorophyta. This relationship is implied because green algae and land plants share a common cell wall chemistry, use the same types of chlorophyll, and also store starch.

Nonvascular Land Plants

Bryophytes are land plants but they require a wet environment in order to reproduce. They are generally small, and like the algae, do not have vascular tissue. The gametophyte generation is the one which you commonly see as the green plant. The sporophyte generation grows out of the gametophyte plant's body, and is often brown or orange.

Vascular Plants

Vascular tissues help the ferns, gymnosperms and angiosperms survive in dry environments, and to grow larger. These groups have true leaves, stems, and roots. Leaves are important for photosynthesis, stems provide support and allow for transport of materials, and roots are used for water absorption and holding the plant in place.

Ferns do not produce seeds but gymnosperms and angiosperms do. Gymnosperm is an old botanical term that no longer represents a taxonomic group, but is widely used as a way of grouping four divisions together: the coniferophyta, cycadophyta, ginkgophyta, and gnetophyta. The conifers account for the largest, oldest, and most successful members of the gymnosperms.

Flowers are only found in the angiosperm, which is currently the dominant plant division on earth. It is possible that they reached this status by their superior ability to adapt to major changes in the earth's climate over the last several million years. Flowers continue to evolve, and four major trends are apparent today. The members of this division also have seeds which are surrounded (and protected) by tissue from the ovary.

KEY TERMS

Be sure to make and use flash cards for all of these terms. Suggestions are found in "To The Student" at the beginning of the Study Guide. Page numbers refer to the text.

algae	319	gymnosperm	327
alternation of generations	317	holdfast	322
angiosperm	331	leaf	326
bryophyte	323	phaeophyte	321
chlorophyte	322	phycoerythrin	320
coniferata	327	pterophyte	326
coniferophyta	319	rhizoid	326
cycadophyta	327	rhizome	326
division	319	rhodophyte	320
frond	326	root	326
fucoxanthin	322	sporophyte	318
gametophyte	318	sporophytic	317
gametophytic	317	stalk	322
gingkophyta	327	thalli	322
gnetophyta	327	vascular	324

EXERCISES

Check your understanding of Chapter 12 by completing the following exercises. Answers begin on p. 207.

1. Use the clues below to complete the puzzle. Write a letter in each blank.

a.

a. ___ ___ ___ ___ ___

b. ___ ___ ___ ___ ___

c. ___ ___ ___ ___ ___ ___

d. ___ ___ ___ ___ ___ ___ ___ ___

e. ___ ___ ___ ___ ___ ___ ___

f. ___ ___ ___ ___

Clues Across
- a. fern leaf
- b. aquatic plants
- c. genus with only one living species
- d. very small, but produces flowers
- e. genus related to cypress
- f. plant anchor

Clue Down
- a. important in reproduction of Anthophyta

True or False Questions

Mark each question either T (True) or F (False).

____2. In plants, spores are produced by the sporophyte generation as a result of meiosis.

____3. In plants, gametes are produced by the gametophyte generation as a result of meiosis.

____4. All land plants produce seeds.

____5. Land plants no longer use fertilization.

Matching Questions

Each question may have more than one answer; write all of the answers that apply. Answers may be used more than once.

____6. is conspicuous in angiosperms

____7. produces spores during reproduction

____8. is haploid

a. gametophyte generation
b. sporophyte generation
c. both of the above
d. none of the above

____9. shows alternation of generations

____10. has flowers

____11. has true roots

____12. produces gametes

____13. contains specialized vascular tissue

a. fern
b. flowering plant
c. bryophyte
d. gymnosperm
e. all of the above
f. none of the above

Multiple-Choice Questions

Circle the choice that is the best answer for each question.

14. The gametes of green algae will be
    a. haploid.
    b. diploid.
    c. both of the above are possible, depending on which generation produces them.

15. The _____ were the most likely ancestors to land plants.
    a. Phaeophyta
    b. Chlorophyta
    c. Bryophyta
    d. Cycadophyta
    e. none of the above

16. Alternation of generations will provide the opportunity for
    a. new alleles to enter the population.
    b. alleles that are in the population to mix in new combinations.
    c. alleles from one division to mix with alleles from a different division.
    d. alleles of nonvascular plants to become alleles of flowering plants.

Fill in the Blanks

Complete each statement by writing the correct word or words in every blank.

17. The red algae have chlorophyll, but they also have the pigment __________, which gives them a __________ color. This pigment helps them accomplish photosynthesis in __________ water.

18. Cycads use the pigments __________ and __________ for photosynthesis.

19. Giant kelps can grow to lengths of __________ meters. They are in the division __________, commonly called the __________ algae.

20. In some species of plants, a period of stress will result in an __________ in reproduction.

21. Conifers store their food as __________.

22. A major problem for land plants was the __________ air.

Questions For Further Thought

Write your answer for each question in the space provided.

23. When you are walking through the woods, how can you tell a large moss from a small conifer? Assume that each is fully mature and reproducing.

SUMMARY

Study the ten items listed in the SUMMARY on p. 333 of the text. Some members of the plant kingdom live as microscopic colonies, yet others are complex multicellular organisms that grow over 300 feet tall. Their tremendous diversity has enabled them to inhabit all regions on the earth. You should have noticed that photosynthesis and alternation of generations occur in all of the divisions, yet each group also has its own unique characteristics which distinguish it from the other groups. Many types of plants seem to have changed very little for long periods of time, however continuing evolution is clearly visible in the flowering plants.

HINTS

Read these suggestions before you check your answers to the EXERCISES.

For question 3:	The gametophyte is already haploid. Can it go through meiosis?
For question 4:	Remember, bryophytes are land plants.
For question 7:	Think about alternation of generations. Does it occur in all plants?
For question 9:	See the hint for Question #.
For question 11:	Question # relates to this.
For question 13:	If you figured out Question #, you have the answer for this one.
For question 14:	This applies to any gametes.
For question16 :	Don't forget what you learned about independent assortment and mutation.
For question 18:	There is more than one kind of chlorophyll.
For question 22:	Is air as wet as water?

CONGRATULATIONS ... YOU HAVE COMPLETED CHAPTER 12 !!!

CHAPTER 13

DIVERSITY III: ANIMALS

CONNECTIONS

Animals are the only kingdom left on our tour through the living things of the earth. Although this is probably the group with which you are most familiar, you should be prepared for some new information and a few surprises. This chapter gives you a brief introduction to the major animal phyla and their characteristics. The details of reproduction and development are in Chapters Fifteen and Sixteen.

At first glance this chapter might just look like a very long list of characteristics to memorize. If that were all there was you could simply learn Table 13.1, which is probably a good idea anyway. In addition, watch for the discussion of evolutionary changes and the ways in which they affect each group of animals. The animal kingdom shows many good examples of the relationship between structure and function. Be sure to pay attention to these throughout the chapter.

OBJECTIVES FOR THIS CHAPTER

When you have finished Chapter 13, you should be able to:

1. Name the major animal phyla and their classes.
2. Describe the body plan of each class.
3. Distinguish between simple and advanced characteristics.
4. Describe the major evolutionary changes seen in each phylum and discuss how they relate to the characteristics of other phyla.
5. Describe two types of body symmetry and relate each one to an animal's lifestyle.
6. State the evolutionary importance of cephalization.
7. Name the features common to all chordates.
8. Prepare an evolutionary tree that shows the origins of each major phyla.

CHAPTER 13 OUTLINE

I. The Animal Kingdom
   A. Phylum Porifera: An Evolutionary Dead End
   B. Phylum Cnidaria and the Evolution of Radial Symmetry
   C. Phylum Platyhelminthes and the Evolution of Heads and Sides
   D. Phylum Nematoda and the Evolution of a Body Cavity
   E. Phylum Mollusca and the Evolution of the Coelom
   F. Phylum Annelida and the Evolution of Repeating Segments
   G. Phylum Arthropoda and the Evolution of Specialized Segments
   H. Phylum Echinodermata and an Evolutionary Puzzle
   I. Phylum Hemichordata and a Half Step Along
   J. Phylum Chordata and the Advent of Freeways
      1. Class Agnatha
      2. Class Chondrichthyes
      3. Class Osteichthyes
      4. Class Amphibia
      5. Class Reptilia
      6. Class Aves
      7. Class Mammalia

CONCEPTS IN BRIEF

The Animal Kingdom

Study Table 13.1 for the basic characteristics of each animal phylum. The following discussion will focus on additional features of the animals.

Sponges seem to be by themselves in the family tree of animals with their own protistan origins and no apparent close relatives. Their specialized cells make them too advanced to be placed in the Kingdom Protista, so by general agreement they are here with the animals.

Adult **cnidarians** have radial symmetry. This is advantageous for animals which do not have a head, since it allows them to sense the environment in all directions. The larvae show bilateral symmetry, and therefore it has been suggested that they may have given rise to the flatworms.

Flatworms have a head at the front end of their body, and move forward through the environment. Having accumulations of sensory and nervous tissues at their leading end (they move forward) gives them the advantage of being able to detect what is in the environment ahead of themselves. Developing top and bottom sides as well as bilateral symmetry may be related to cephalization as well. The tapeworms have become so specialized as parasites that they have become more simplified through time.

Nematodes have a two-opening gut that is surrounded by a body cavity. This allows a place for the development of internal organs, and separates the actions of the muscles involved with movement from those involved in digestion. Although this body cavity is a pseudocoelom (because of the distribution of muscle tissue) it functions like a true coelom.

The rest of the animal phyla have true coeloms. In addition, they are divided into groups of phyla known as the protostomes and deuterostomes, depending on which embryonic opening becomes the mouth and which becomes the anus.

Mollusks are quite diverse, but they all have a muscular foot (which have developed into tentacles in some species) and a mantle which covers the soft body.

Annelids show the specializations of repeated body parts in segments and a circulatory system which has blood inside vessels. These allow for more efficient functioning and greater control in a large animal body.

Arthropods have been extremely successful. They make use of specialized body segments and a complex nervous system. Jointed legs give them mobility through many types of environments, and their exoskeleton provides protection.

Echinoderms are known for their efficient and powerful water vascular system, which is used to assist in movement. Many of them can push their stomach out of their body and turn it inside out while digesting prey that is too big to "swallow." Their radial symmetry is unique among the deuterostomes.

Chordates are divided into three subphyla but all of them have three characteristics in common at some time in their lives: (1) a notochord, (2) a dorsal hollow nerve cord with a brain at the front end, and (3) gill slits.

Vertebrates are the subphylum of chordates to which humans and most zoo animals belong. They are distinguished from other chordates by having the dorsal hollow nerve cord protected by vertebrae.

Jawless fishes are either parasitic on other animals or are scavengers. Individuals have both male and female reproductive organs.

Chordates with **jaws** are, of course, able to develop styles of feeding and relationships with other animals that are not seen in the Agnatha.

Cartilaginous fish have skeletons made of cartilage rather than true bone, as in the **bony fishes**. See how easy some of this terminology can be?

Amphibians, as a group, have made the successful transition to life on land. They still rely on a watery environment for fertilization and for their developing eggs and larvae.

Reptiles use internal fertilization and other characteristics which have allowed them to be successful on land. Ancient reptiles evolved into modern reptiles, birds, and mammals.

Birds use feathers and internal physiological controls to regulate their body temperature. They have many features which, in combination, enable most of them to fly.

Mammals also have very good body temperature controls and produce milk for their young. You are in the class Mammalia, and the order Primates. A great deal of the rest of the text book will be devoted to the activities of mammals.

KEY TERMS

Be sure to make and use flash cards for all of these terms. Suggestions are found in "To The Student" at the beginning of the Study Guide. Page numbers refer to the text.

abdomen	355	foot	349
amoebocyte	339	gastrodermis	340
annelid	350	gastrovascular cavity	340
Arthropoda	352	hemichordate	357
cephalization	344	hermaphroditic	360
chordate	357	mantle	349
clitellum	351	medusa	341
Cnidaria	340	mesoglea	340
coelom	347	metazoan	337
collar cell	339	mollusk	347
deuterostome	346	nematode	346
echinoderm	356	notochord	357
epidermis	340	parapodia	352

EXERCISES

Check your understanding of Chapter 13 by completing the following exercises. Answers begin on p. 207.

1. Use the following terms to label the diagram (of a typical vertebrate) below.

   anus/cloacal opening, brain, coelom, digestive tract, gill slits, mouth, nerve cord, vertebrae

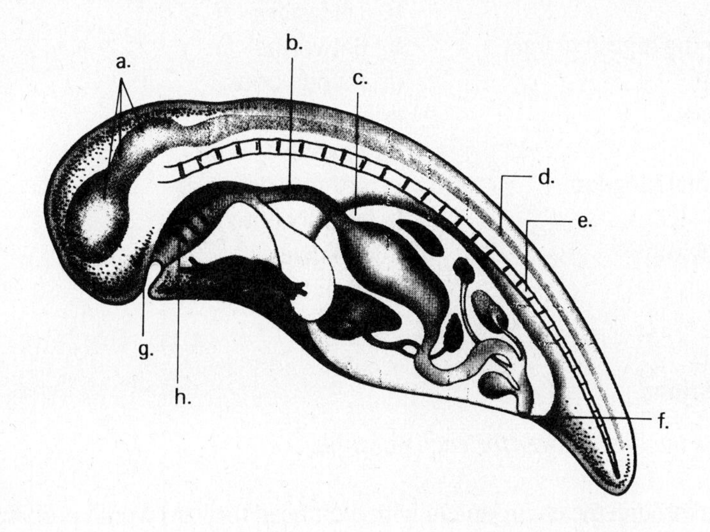

2. Classify yourself by filling in the blanks.

   Kingdom ____________________

   Phylum ____________________

   Class ____________________

   Order ____________________

True or False Questions

Mark each question either T (True) or F (False).

____3. Flatworms show cephalization.

____4. Sponges have specialized cells.

____5. An earthworm can use its male organs to fertilize its own female organs.

____6. The skeleton of an insect is made of cartilage.

____7. Polychaetes have repeating parts in their bodies.

Matching Questions

Each question may have more than one answer; write all of the answers that apply. Answers may be used more than once.

____8. adults have bilateral symmetry

____9. use a two-opening digestive tract

____10. are protostomes

____11. are in the animal kingdom

____12. have a true coelom

____13. have true nerves

a. sponges
b. cnidarians
c. flatworms
d. roundworms
e. mollusks
f. annelids
g. arthropods
h. echinoderms
i. chordates
j. all of the above

Multiple-Choice Questions

Circle the choice that is the best answer for each question.

14. Animals that move through the environment with their head forward would probably have _____ symmetry.
    a. no
    b. bilateral
    c. radial
    d. metazoan

15. Echinoderms are placed close to chordates on an evolutionary tree because
    a. they have a similar ratio of brain weight to body weight.
    b. each has a gastrovascular cavity.
    c. they show the same style of development as an embryo.
    d. the vertebral column of each shows the same pattern.

16. The closed circulatory system is first seen in the
    a. roundworms.

b. annelids.
c. arthropods.
d. chordates.

17. Vertebrates probably came from ancestral
  a. echinoderms.
  b. flatworms.
  c. cephalochordates.
  d. urochordates.

Fill in the Blanks

Complete each statement by writing the correct word or words in every blank.

18. In the development of the embryo of protostomes, the first opening becomes the _________.

19. Elephants have _________ symmetry.

20. A nerve net is found in the _________.

21. Mollusks have a _________ body which is covered by a _________.

Questions For Further Thought

Write your answer for each question in the space provided.

22. Amphibians were the first successful land vertebrates. Discuss the unique problems they were faced with and the possible benefits.

SUMMARY

Study the twenty items listed in the SUMMARY on p. 375 of the text. As you read about the different phyla it should have become clear that there are many successful ways to solve the same problems in the animal kingdom. Several evolutionary changes can be seen that gave various groups an advantage in the environment. Many of these changes are then seen in other phyla as you move up the evolutionary tree. Remember that the animals alive on the earth today <u>are</u> successful in their own environments with their own sets of characteristics, even if they seem to be evolutionary dead ends.

HINTS

Read these suggestions before you check your answers to the EXERCISES.

For question 4:	The key word here is "cells."
For question 6:	Don't confuse insects with sharks.
For question 7:	Think about segmentation.
For question 15:	Think about protostomes and deuterostomes.
For question 17:	This asks about vertebrates, not chordates.
For question 19:	Elephants are vertebrates.
For question 21:	The first answer is either "hard" or "soft."

CONGRATULATIONS ... YOU HAVE COMPLETED CHAPTER 13 !!!

Part Four Reproduction and Development

CHAPTER 14

PLANT REPRODUCTION AND DEVELOPMENT

CONNECTIONS

Your focus during Part Four of the text will be on the plants and animals. By now you are familiar with the characteristics of the major groups of organisms in each of these kingdoms. The coming chapters will present the details of mating, growth and development. As always, look for common themes or problems and the variety of ways they may be solved.

In the current chapter we return to alternation of generations in plants with examples from selected plant divisions. Watch for the ways that sexual and asexual reproduction affect plant populations and their evolution. Remember, plants have rigid cell walls. Pay attention to the unique problems which this presents to plants while they grow.

OBJECTIVES FOR THIS CHAPTER

When you have finished Chapter 14, you should be able to:

1. Differentiate between sexual and asexual reproduction in plants.
2. Discuss the advantages and disadvantages of sexual and asexual reproduction in plants.
3. Use examples, and describe the reproductive cycle of each of the following: include both the gametophyte and sporophyte generations.

   green algae
   mosses

ferns
conifers
flowering plants

4. Outline the process of seed production in conifers and flowering plants.
5. List the structures of a typical angiosperm flower.
6. Discuss the features which made it possible for large plants to be successful in dry land environments.
7. Name and describe the plant tissues that are produced by differentiation.
8. Describe the functions of the apical and lateral meristems.
9. Name three plant hormones and state how they influence plant growth and development.

CHAPTER 14 OUTLINE

I. Sexual Reproduction
   A. Disadvantages of Sexual Reproduction
   B. Advantages of Sexual Reproduction
II. Reproduction and Life Cycles in Various Plants
   A. Green Algae
   B. Mosses
   C. Ferns
   D. Conifers
   E. Flowering Plants
   F. How Flowers Work
III. Plant Development
IV. Plant Hormones
   A. Auxins
   B. Gibberellins
   C. Cytokinins

CONCEPTS IN BRIEF

Sexual Reproduction

Variation is a key issue in this section. Remember that variation is beneficial in a population that must adapt to a changing environment, and both meiosis and sexual reproduction lead to an increase in variation. On the other hand, asexual reproduction assures that offspring will have successful genotypes in a stable environment. Of course organisms can't predict what the environment will do in the future and choose their reproductive style accordingly. The apparent disadvantages that sexual reproduction presents to individual plants may be outweighed by the possible long term advantages which the population receives.

Reproduction and Life Cycles in Various Plants

If you are a bit fuzzy on the details of alternation of generations, you should return to Chapter Twelve to refresh your memory. The haploid (gametophyte) generation is the dominant (or most conspicuous) generation in mosses and most green algae. In ferns, conifers, and flowering plants the diploid (sporophyte) generation is dominant. These divisions with dominant sporophytes also have very small gametophytes. In the conifers and angiosperms the gametophyte is microscopic and develops inside (and is protected by) the sporophyte plant.

Plants commonly produce resistant structures which provide protection during periods of stress. Examples of these include: the zygospores of green algae, moss spores, and both pollen and seeds of conifers and angiosperms.

Fertilization in the green algae, mosses, and ferns occurs after the sperm swim (at least through a film of water) to the egg. In some complex green algae we find the first examples of large immobile eggs that are located by the sperm. The pollen of conifers and angiosperms is able to survive as it moves through the dry air. This allows fertilization to occur without an external film of water.

After fertilization occurs, a zygote (the beginning of the sporophyte generation) is produced. In mosses and ferns, this sporophyte plant grows directly out of the gametophyte plant. The moss gametophyte remains in place with the sporophyte on top of it, but the fern gametophyte eventually withers away. The immature sporophyte of both the conifers and flowering plants becomes dormant and is encased in the seed, which offers some protection from the environment. In the flowering plants the seed itself is also enclosed in a fruit.

Plant Development

Plants continue to grow throughout their lives. Because of their rigid cell walls, plants grow by adding new cells to their outer layers and growing tips. Older cells are either surrounded or left behind. This section of the text only deals with development in flowering plants.

As seeds begin to grow, the cells take on their specialized roles and all of the body parts of the plant are eventually produced. The main body regions of the plant are: outer protective areas (epidermis), vascular tissue which moves food and water inside the plant, and the remaining bulk of tissue (cortex) which provides support. Growth in roots and stems occurs in the plant's tips due to cell division by the apical meristem tissue. The plants grow broader because of cell division in one or more layers of cambium. Cambium must also produce new vascular tissue and cork in the outer layers of the plant.

Plant Hormones

Plants have numerous hormones which regulate a wide variety of functions. The three groups of hormones considered here are the auxins, gibberellins, and cytokinins. Table 14.1 presents a concise view of their actions and effects. A great deal of work still needs to be done in order for scientists to understand the hormonal systems of plants.

KEY TERMS

Be sure to make and use flash cards for all of these terms. Suggestions are found in "To The Student" at the beginning of the Study Guide. Page numbers refer to the text.

alternation of generations	384	carpel	392
anther	392	collenchyma	398
antheridium	386	cork	398
apical meristem	396	cork cambium	398
archegonium	386	cortex	396
auxin	400	cotyledon	396
bud scale	396	cytokinin	402
cambium	397	dicot	396

dicotyledon	396
differentiation	395
egg nucleus	389
embryo	395
embryo sac	392
endosperm	393
epidermis	395
filament	392
fruit	394
gametophyte	384
generative cell	392
genetic imprinting	384
gibberellin	401
hormone	398
lateral meristem	397
megaspore mother cell	388
micropyle	392
microspore mother cell	388
monocot	396
monocotyledon	396
ovary	392
ovule	392
parenchyma	398
petal	390
phloem	397
pistillate	388
polar nuclei	392
pollen	392
pollen grain	392
pollen sac	392
pollen tube	390
pollen tube cell	392
receptacle	390
sclerenchyma	398
seed	390
seed coat	393
sepal	392
sperm nucleu	s390
sporophyte	384
stamen	392
staminate	388
stigma	392
style	392
terminal bud	396
vascular cambium	398
vascular tissue	396
xylem	397
zygote	391

EXERCISES

Check your understanding of Chapter 14 by completing the following exercises. Answers begin on p. 207.

1. Label the following diagram of a flower.

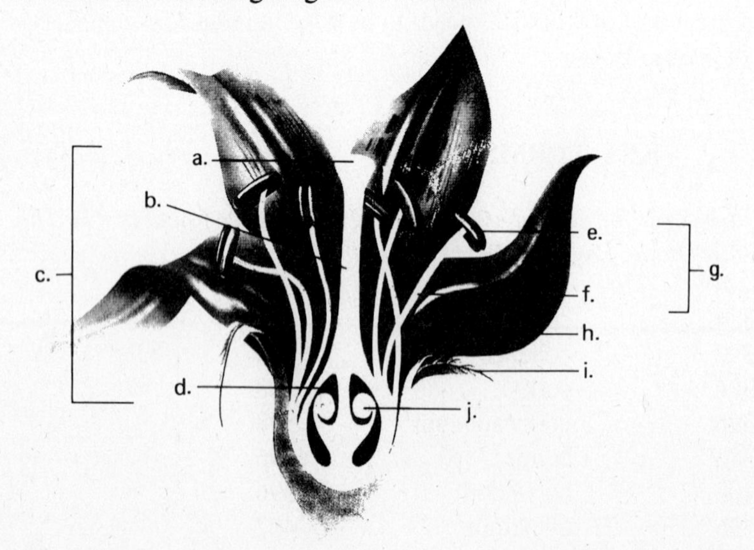

a._______
b._______
c._______
d._______
e._______
f._______
g._______
h._______
i._______
j._______

2. Use the diagram from Question # 1 to answer the following items. Write the proper letter or letters in each blank.

_____ a. site of pollen formation

_____ b. where pollen lands during pollination

_____ c. site of fertilization

_____ d. pollen tube grows here

_____ e. develops into a fruit

_____ f. location of megaspores

_____ g. contains at least some haploid nuclei

True or False Questions

Mark each question either T (True) or F (False).

_____3. If you cut the tip off an oat seedling, apply auxin to one side of what is now the tip, and put it in the dark, the seedling will bend away from the side with the auxin.

_____4. A conifer embryo is found inside the seed.

_____5. Microspore mother cells undergo meiosis in the beginning stages of egg formation in angiosperms.

_____6. Ferns require a film of water on the gametophyte for the sperm to swim to the egg.

_____7. The sporophyte generation is dominant in mosses.

_____8. An increase in variation is one result of asexual reproduction.

Matching Questions

Each question may have more than one answer; write all of the answers that apply. Answers may be used more than once.

_____9. produces new cells

_____10. transports water

_____11. acts as a protective outer region

_____12. is composed of diploid cells

a. epidermis
b. xylem
c. phloem
d. apical meristem
e. cambium
f. all of the above
g. none of the above

Multiple-Choice Questions

Circle the choice that is the best answer for each question.

13. Plant populations which live in a changing environment are more likely to survive if they are able to
    a. reproduce sexually.
    b. reproduce asexually.
    c. keep from reproducing at all.
    d. suppress meiosis before times of environmental change.

14. If you were to look from mosses to ferns to conifers, which of the following trends would you notice?
    a. Meiosis becomes less important.
    b. Production of spores disappears.
    c. Gametophytes become less visible.
    d. The haploid stage grows taller.
    e. All of the above.

15. Angiosperms grow in width by adding new cells
    a. to the center of the plant.
    b. to the outer regions of the plant.
    c. throughout all parts of the plant.
    d. only in the buds.

16. Cell division is stimulated by
    a. auxins.
    b. gibberellins.
    c. the micropyle.
    d. sclerenchyma cells.
    e. all of the above.

17. The gametes of ferns are
    a. always haploid.
    b. always diploid.
    c. diploid only under dry conditions.
    d. only haploid if they are produced by the sporophyte.
    e. none of the above.

Fill in the Blanks

Complete each statement by writing the correct word or words in every blank.

18. During conifer reproduction, each spore will develop into a(n) __________ __________.

19. Apical meristems are found in the __________ tips and __________ tips of flowering plants. They help the plant grow __________.

20. If gibberellins are applied to the stem of a flowering plant, the stems will grow _________.

21. In times of stress, green algae may produce _________ and go through _________ reproduction.

22. The weed killer 2,4,5-T is a synthetic form of _________.

Questions For Further Thought

Write your answer for each question in the space provided.

23. a. Describe the kind of environment that is best for plants which use asexual reproduction.

b. Discuss why you chose *this* environment.

24. Explain how pollen and seeds were important in helping plants become successful on land.

SUMMARY

Study the sixteen items listed in the SUMMARY on p. 403 of the text. You have seen that under different circumstances both asexual and sexual reproduction can be advantageous. Each plant division has its own specialized reproductive structures and patterns of development. Survival of large plants in a wide range of dry land environments has been possible due to pollen seeds, and a dominant sporophyte generation with its smaller, protected gametophyte generation. In general, plants are much more active and complex than they may appear to be at first glance.

HINTS

Read these suggestions before you check your answers to the EXERCISES.

For question 2:	Remember when and where meiosis occurs.
For question 5:"	Micro" is usually associated with males, "macro" with females.
For question 8.:	Asexual reproduction involves mitosis.
For question 14.:	Don't forget about alternation of generations. It applies to all plants!
For question 15:	Remember, plant cells have rigid cell walls.
For question 16:	The key word here is "division."
For question 17:	Actually, this question applies to all the plants, not just ferns.

CONGRATULATIONS . . . YOU HAVE COMPLETED CHAPTER 14 !!!

CHAPTER 15

ANIMAL REPRODUCTION

CONNECTIONS

As you think back on the earlier chapters of the text you will remember that successful reproduction (and lots of it) is the key to long term survival of a population. In Chapter Thirteen you may have noticed that methods of animal reproduction were only mentioned briefly in one of the tables. Now it is time for a more detailed look at the various ways in which animals reproduce. The emphasis in this chapter is on humans, as a representative example of both a mammal and of the process internal fertilization.

You probably (and understandably) think of sex when you think of animal reproduction. There are, however, several types of animals that are able to reproduce asexually. Look for the advantages and disadvantages of these methods of reproduction for animals. This should remind you of your study of reproduction in plants (as first described in Chapter Twelve).

Be sure to relate birth control techniques and their effectiveness to male and female anatomy and physiology. Also look for connections between birth control and human reproductive behavior.

OBJECTIVES FOR THIS CHAPTER

When you have finished Chapter 15, you should be able to:

1. Describe ways in which animals reproduce asexually.
2. Explain the advantages and disadvantages of asexual and sexual reproduction in animals.
3. Use examples and describe the reproductive strategies involved in internal and external fertilization.
4. List the reproductive structures of the human male and female and describe the functions of each.
5. Explain how the menstrual cycle is regulated.
6. Describe the physiological changes that occur in males and females during each stage of the human sexual response.
7. Describe how conception occurs.
8. List the various means of birth control and their relative effectiveness in preventing pregnancy, implantation, or birth.

CHAPTER 15 OUTLINE

I. Asexual Reproduction
  A. Binary Fission
  B. Regeneration
  C. Budding
II. Sexual Reproduction
  A. Timing of Sexual Reproduction
III. Human Reproduction
  A. The Male Reproductive System
  B. The Female Reproductive System
    1. The Menstrual Cycle
IV. The Human Sexual Response
  A. Conception
  B. Sex and Society
V. Contraception
  A. Historical Methods
  B. The Rhythm Method
  C. Coitus Interruptus
  D. The Condom
  E. The Diaphragm
  F. The Cervical Cap
  G. Spermicides
  H. The Intrauterine Device
  I. The Birth Control Pill
  J. Norplant
  K. The Sponge
  L. Sterilization
  M. Abortion

CONCEPTS IN BRIEF

Asexual Reproduction

Both asexual and sexual reproduction can occur in the sponges, cnidarians, and flatworms. As in plants, asexual reproduction produces offspring with the same genotype as the parent, which may be beneficial in a stable environment.

Asexual reproduction allows the production of offspring to occur when there are no potential mates to be found. In addition, if these animals are injured they can repair their parts. In some cases the lost parts can develop into a complete new individual.

Sexual Reproduction

Sexual reproduction occurs when an egg is fertilized by a sperm. The zygote which is produced can grow to be a new individual. In external fertilization, large numbers of eggs and sperm are released into the environment. Successful fertilization depends on: timing, location, the ability of the gametes to survive on their own long enough to find one another and achieve fertilization, and a lot of chance.

The development of internal fertilization allows organisms to produce fewer gametes and to keep them inside their bodies, where there is a less hostile environment. The organisms must still find a mate and copulate at a time (which may be regulated by the environment) when both eggs and sperm are ready in order to have a chance of producing offspring. It is common for females to signal males when they are receptive and fertile.

Human Reproduction

The male reproductive system is designed to produce large numbers of sperm (at appropriately low body temperatures) and to deposit them into the vagina of the female. The female's system receives the sperm and provides the location for fertilization and subsequent development of the zygote. Tables 15.1 and 15.2 as well as Figures 15.9 and 15.10 supply the structural and functional details for males and females.

From puberty through menopause, a woman's body repeatedly prepares her reproductive tract to release a fertile egg from an ovary, and to maintain any developing zygote (in the case of successful fertilization) in the uterus. The hormones FSH, LH, estrogen, and progesterone interact to regulate these events in what is known as the menstrual cycle. If pregnancy does not occur, the built up endometrium of the uterus is sloughed off and expelled as menstrual flow, and the cycle begins again.

The Human Sexual Response

Humans attach varying levels of emotion to all aspects of sex and sexuality. The text explores various implications of these feelings and their relationship to successful reproduction and pair bonding.

The activities of foreplay prepare both the male and the female for the events of copulation. In particular, the penis becomes erect and the vagina is lubricated. Orgasm in men involves contractions of the muscles along the reproductive tract followed by ejaculation, which forces the semen out. Female orgasm also involves rhythmic muscle contractions, but its physiology is not clearly understood.

Sperm must travel from the vagina, through the cervix and uterus and into the oviduct where it will find the egg, and fertilization can occur. The resulting zygote will travel to the uterus where it will implant and continue to develop.

Contraception

Many effective and ineffective techniques for preventing pregnancy and birth have been used in the past or are in use today. Choosing a method requires the examination of advantages and disadvantages, personal preferences and beliefs, effectiveness, availability, and cost.

The rhythm method and coitus interruptus involve regulating one's behavior to prevent sperm from being present when an egg is available.

Placing a physical barrier in the way of the sperm is the concept behind condoms, diaphragms, and cervical caps. Spermicide is often used with these, or it may be used alone. Sponges are only effective because of the spermicide in them.

Birth control pills and Norplant rods rely on hormones which prevent eggs from developing and being released from the ovary.

Sterilization of males is simpler to do than is sterilization of females. Either method is very effective, but there is always the possibility that the sterility will be permanent.

"Morning after pills" and use of the IUD are methods which prevent implantation. The IUD has fallen out of favor in the United States due to its side effects. Abortions are used to end a pregnancy. Legal abortions are much safer than illegal procedures, but the entire issue is very clouded by emotional arguments.

KEY TERMS

Be sure to make and use flash cards for all of these terms. Suggestions are found in "To The Student" at the beginning of the Study Guide. Page numbers refer to the text.

abortion	434	labia majora	418
binary fission	408	labia minora	418
birth control pill	431	luteinizing hormone (LH)	419
budding	410	menstruation	418
bulbourethral glands		orgasm	422
(Cowper's glands)	415	ovaries	417
cervical cap	430	oviduct	417
cervix	417	ovulation	419
clitoris	418	penis	417
coitus interruptus	427	progesterone	419
conception	423	prostate gland	415
condom	427	regeneration	410
contraception	425	rhythm method	427
copulation	413	scrotum	415
corpus luteum	419	seminal vesicles	415
diaphragm	430	seminiferous tubules	415
ejaculation	415	spermicide	430
endometrium	417	sponge	432
epididymis	415	sterilization	432
estrogen	419	testes	415
estrus	413	tubal ligation	433
fallopian plug	434	urethra	415
fertilization	423	uterus	417
follicle	419	vagina	418
follicle-stimulating		vas deferens	415
hormone (FSH)	419	vasectomy	432
intrauterine device (IUD)	430	vulva	418

EXERCISES

Check your understanding of Chapter 15 by completing the following exercises. Answers begin on p. 207.

1. Fill in the names of the structures on the following diagrams.

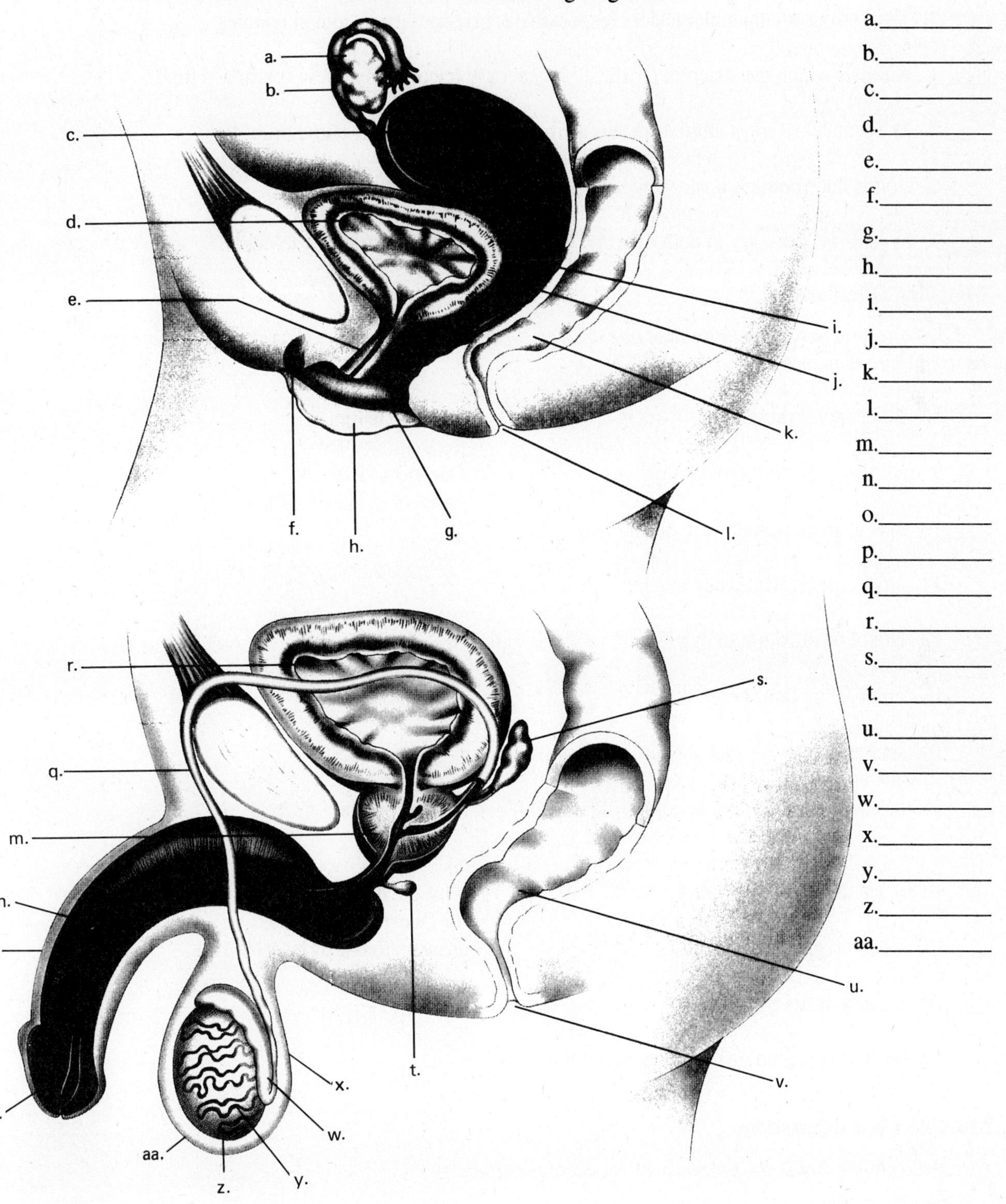

True or False Questions

Mark each question either T (True) or F (False).

____2. Binary fission is a form of asexual reproduction.

____3. The seasons are the main factors responsible for receptivity in human females.

____4. Animals which use external fertilization generally release only a few sperm at a time.

____5. The females of most animal species are receptive to mating all of the time.

____6. Coitus interruptus is a very effective method of conception control.

____7. Orgasm is necessary in both men and women in order for pregnancy to occur.

Matching Questions

Each question may have more than one answer; write all of the answers that apply. Answers may be used more than once.

(Use the letters in the diagrams for Question #1 for the answers to questions #8–15.)

____8. site of sperm production

____9. releases eggs during ovulation

____10. sperm are deposited here during intercourse

____11. stores sperm while they mature

____12. site of fertilization of egg

____13. embryo implants here

____14. secretes materials that are part of the semen

____15. has spongy tissue that becomes filled with blood during foreplay

(Use the letters in the diagrams for Question #1 for the answers to questions #16–19.)

____16. is cut during a vasectomy

____17. a diaphragm fits over this

____19. a condom fits over this

____20. an IUD is placed inside the cavity of this

Multiple-Choice Questions

Circle the choice that is the best answer for each question.

20. Regeneration of lost body parts is common in

a. sea stars.
b. bedbugs.
c. chimpanzees.
d. mammals.
d. all of the above.

21. Asexual reproduction in animals can be accomplished by
a. ovulation.
b. budding.
c. copulation.
d. menstruation.
e. all of the above.

22. The _____ secretes LH, which triggers _____.
a. ovary, ovulation
b. corpus luteum, menstruation
c. pituitary gland, ovulation
d. uterus, menstruation

23. The most commonly transmitted bacterial disease in America is
a. syphilis.
b. gonorrhea.
c. chlamydia.
d. AIDS.

Fill in the Blanks

Complete each statement by writing the correct word or words in every blank.

24. Frogs use __________ fertilization, in which the sperm are deposited on the __________ in the __________.

25. In humans, the hormone __________ causes the endometrial lining of the __________ to thicken, thereby preparing it for implantation of a(n) __________ egg.

26. Sperm may be stopped by the __________ environment of the vagina, but they survive better in the __________ environment of the uterus.

Questions For Further Thought

Write your answer for each question in the space provided.

27. Assume that you could design the perfect contraceptive. Write a description of your product and how it would be used.

SUMMARY

Study the fifteen items listed in the SUMMARY on p. 435 of the text. You have seen that if organisms are able to find mates, sexual reproduction with internal fertilization probably increases the chances of making offspring. The continual process of human reproduction (and sometimes its prevention) involves a complex interaction of physiological and emotional issues beyond the union of egg and sperm.

Keep in mind that in biological terms, successful reproduction with members of one's own species is essential for the continuation and evolution of the species. The development of complex reproductive systems is intended to ensure that the goal is reached.

HINTS

Read these suggestions before you check your answers to the EXERCISES.

For question 3: This asks about the main cause, not just a possible influence.

For question 14: Are sperm part of the semen?

For question 16: What does the surgeon have to cut through in order to get inside?

For question 17: This assumes the diaphragm is inserted into a female and is positioned correctly.

For question 18: See the hint for Question #, except that this is for a male.

For question 19: In case you haven't caught on to the system yet, see the hint for Question #!

For question 24: The first answer is either "external" or "internal."

For question 25: This is in a female.

For question 26: Think about pH for these answers.

CONGRATULATIONS . . . YOU HAVE COMPLETED CHAPTER 15 !!!

CHAPTER 16

ANIMAL DEVELOPMENT

CONNECTIONS

In the last chapter you studied many aspects of human reproduction, with only a brief look at the events following fertilization. This chapter presents a detailed view of the changes which take place during the development of a typical animal offspring, from the zygote through the point at which it is capable of surviving on its own in the environment.

Although the emphasis is once again on humans, you will also follow the development of frogs and chickens. This will give you examples of three organisms which have very different types of eggs. Watch for the similarities and the differences which are apparent in their embryonic stages.

The next section of your text (Part Five) examines the major body systems of animals. Chapter Sixteen prepares you for this by showing how each of these systems develops. As you examine each system, look for its origin as well as its relationship to the rest of the animal body.

OBJECTIVES FOR THIS CHAPTER

When you have finished Chapter 16, you should be able to:

1. Describe in general how an embryo develops, listing events of fertilization, cleavage, and gastrulation.
2. Name three embryonic germ layers and the tissues formed from each.
3. Describe four ways in which embryonic development is regulated.
4. Explain how the amount of yolk in an organism's egg correlates to its life history.
5. Fully describe the early developmental stages of a frog and a chick.
6. Describe human developmental events during each of the three trimesters.
7. Describe some of the problems which may arise during human development, and their consequences.
8. Relate the sequence of events which occur during birth.

CHAPTER 16 OUTLINE

I. Early Stages of the Embryo

II. Influences on Organization
 A. The Flexible Fate
 B. Sequencing
 C. Stress
 D. Induction
III. Types of Eggs
IV. Early Development in the Frog
 A. The First Cell Divisions
 B. Gastrulation
 C. Embryonic Regulation
V. Birds and Their Membranes
VI. Human Development
 A. The First Trimester
  1. The First Month
  2. The Second Month
  3. The Third Month
 B. The Second Trimester
  1. The Fourth and Fifth Months
  2. The Sixth Month
 C. The Third Trimester
 D. Birth
 E. Miscarriage

CONCEPTS IN BRIEF

Early Stages of the Embryo

Fertilization in animals produces a diploid zygote which begins to divide rapidly, using mitosis with cytokinesis. At a very early stage, these new cells begin to migrate within the embryo and differentiate. Three distinct layers become apparent: the ectoderm, mesoderm, and endoderm. It is possible to predict which adult body structures will develop from each of these layers. The supporting notochord, the early central nervous system, and blocks of muscle tissue soon appear.

Influences on Organization

Many areas of an adult salamander's body are able to regenerate lost tissue because they contain undifferentiated (flexible) cells. In adult mammals most cells follow very specialized roles, and, except for the cells involved in wound healing, are not able to change.

The sequence of events in development is very important in order for the new body parts to coordinate with one another. As you remember, genes on the chromosomes regulate metabolic activities in all cells. Organizing the early events of the embryo is accomplished by genes turning on and off at particular times.

The physical stress of bending and twisting can stimulate tissue development in the embryo. In addition, the presence of one tissue can cause surrounding tissues to develop in certain ways. This can be demonstrated by transplanting embryonic tissues and monitoring the results.

Types of Eggs

Different species of animals produce either small, medium, or large amounts of yolk in their eggs. This seems to relate to their life histories, specifically how long the egg must survive on its own.

Early Development in the Frog

Cleavage in the frog egg results in smaller cells at the animal pole and larger cells at the vegetal pole during the blastula stage. Gastrulation produces the digestive cavity (which opens to the outside via the blastopore) and all three germ layers by the time the embryo is two days old. Some cell differentiation has also occurred by this time.

Birds and Their Membranes

The zygote and its area of cell division occupy a small area (the germinal spot) in one region of a chicken egg. The cells are arranged in a more flattened and linear fashion than in the frog. Again, gastrulation forms all three germ layers. The various membranes perform many important functions: (1) the yolk sac passes food from the yolk, (2) the allantois holds nitrogenous wastes, (3) the amnion and chorion surround everything and the fluids they hold act as shock absorbers. Later, these last two form the chorioallantoic membrane which exchanges oxygen and carbon dioxide.

Human Development

During the first trimester, the development of a human embryo can be altered by many factors. The results may be serious enough to cause birth defects and miscarriage, however if it survives, the developing baby becomes more resilient as it ages.

Early cell division of the embryo, as it moves down the oviduct, is slower than in the frog and chicken. The blastocyst implants in the uterus by the time it is a week old. A placenta forms (from the chorion and the uterine tissue) and acts as a site of nutrient, gas, and waste exchange between the embryo/fetus and the mother until birth.

By the end of the first trimester all of the organ systems have been formed, but they have not completed their development. Bone is present, gender is determined, the heart beats, movement may occur, body features are evident, and it is called a fetus.

The second trimester is the time of much movement on the part of the fetus. It becomes sensitive to outside stimuli, and grows rapidly. The organ systems are not yet mature.

During the third trimester the fetus becomes very crowded as it grows larger. Heavy demands are placed on the mother for food and oxygen. All organ systems become mature and functional, and the mother's antibodies provide a measure of disease resistance.

After the birth of the baby, the placenta is also delivered, which leaves baby to get oxygen from the air and food from nursing.

KEY TERMS

e sure to make and use flash cards for all of these terms. Suggestions are found in "To The Student" at the beginning of the Study Guide. Page numbers refer to the text.

afterbirth	468	chorioallantoic membrane	452
allantois	452	chorion	452
amnion	452	chorionic villi	463
animal pole	447	cleavage	440
archenteron	447	ectoderm	441
blastocoel	447	endoderm	441
blastocyst	463	fertilization	440
blastodisc	451	fetus	464
blastopore	447	gastrulation	441
blastula	447	germ layer	441

germinal spot	451	neural groove	441
inducer	445	neural plate	441
induction	445	notochord	441
inner cell mass	463	placenta	463
labor	467	trophoblast	463
mesenchyme	445	vegetal pole	447
mesoderm	441	yolk	446
morula	447	yolk sac	452
neural folds	441	zygote	440

EXERCISES

Check your understanding of Chapter 16 by completing the following exercises. Answers begin on p. 207.

1. The following diagrams represent stages of development in a frog. Fill in the "a" blanks with the name of each stage. Write numbers in the "b" blanks to indicate the sequence in which they would occur, with "1" being the first stage.

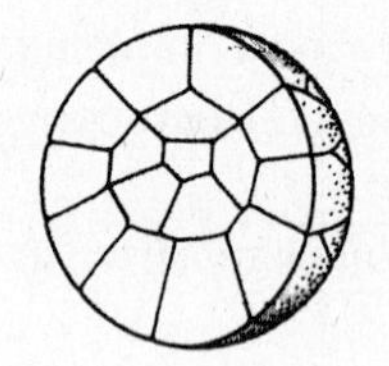
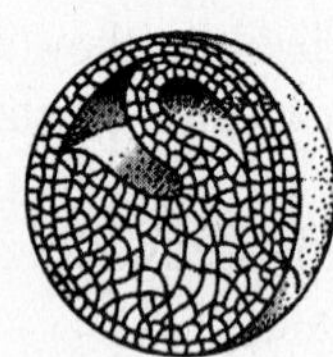
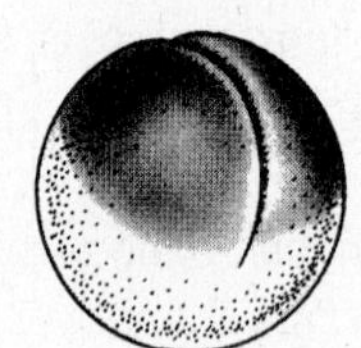
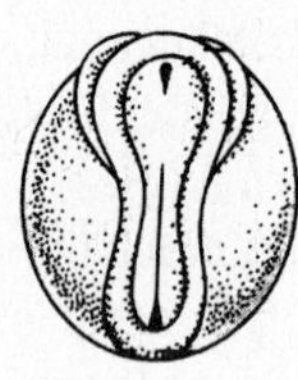
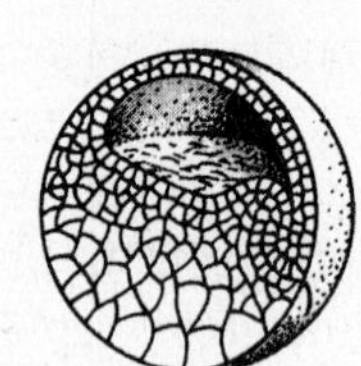
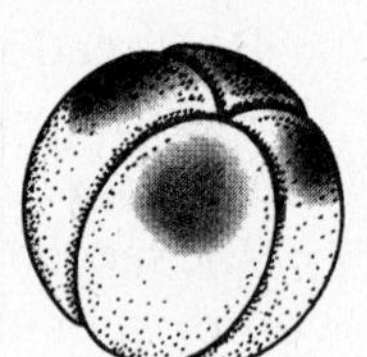

a. _____ _____ _____ _____ _____ _____

b. _____ _____ _____ _____ _____ _____

True or False Questions

Mark each question either T (True) or F (False).

_____2. The zygote stage is found in the development of frogs, chickens, and humans.

_____3. In humans, the morula is a hollow ball of cells.

_____4. Pharyngeal gill slits are found in frog embryos but not in chicken or human embryos.

_____5. If notochord tissue is transplanted at an early enough stage, it can cause spinal cord tissue to develop in parts of the embryo where it would not normally occur.

_____6. The human placenta is reabsorbed by the uterine tissue shortly after the birth of the baby.

Matching Questions

Each question may have more than one answer; write all of the answers that apply. Answers may be used more than once.

Use the letters on the diagram as the answers to Questions 7-13.

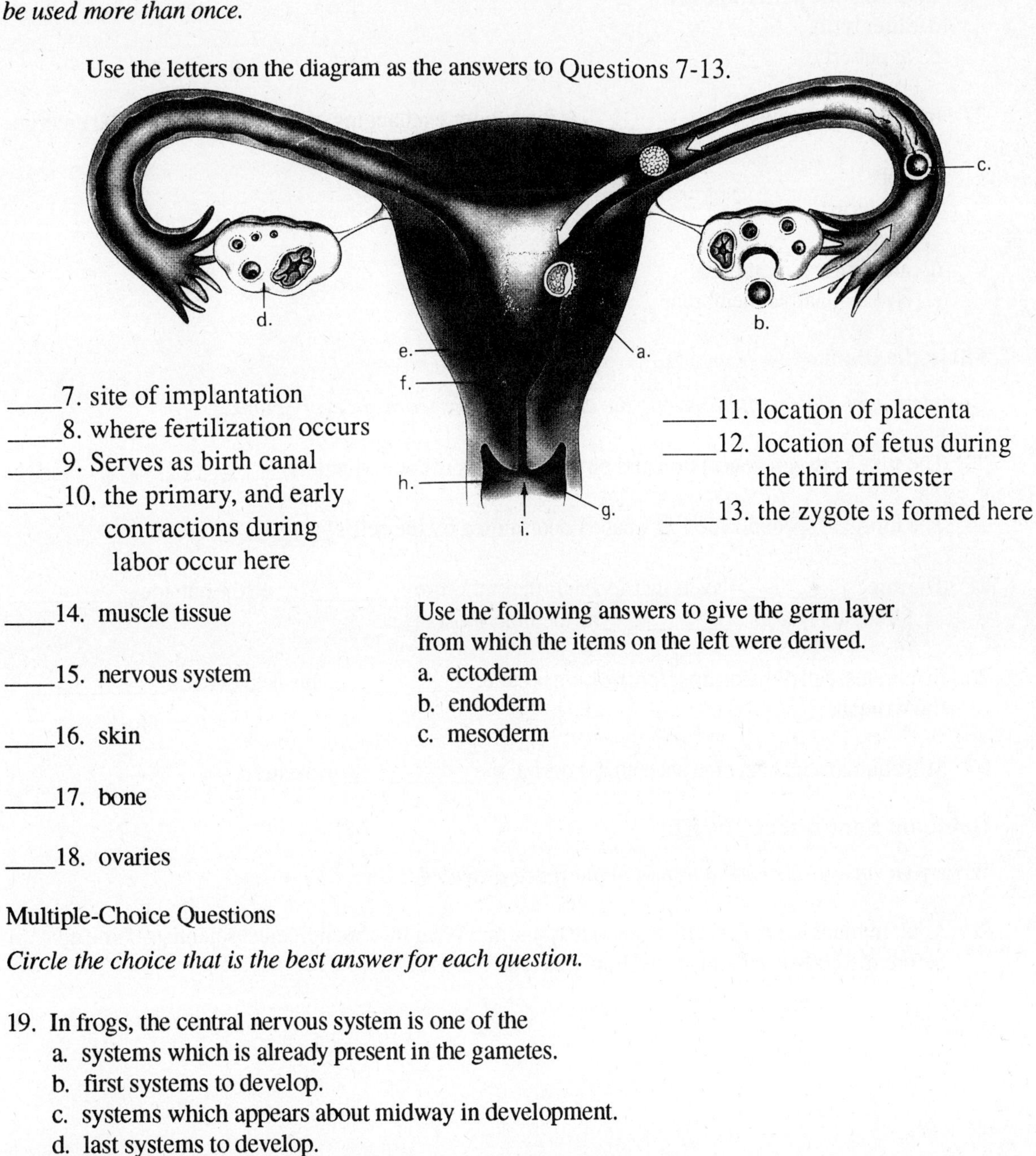

_____7. site of implantation
_____8. where fertilization occurs
_____9. Serves as birth canal
_____10. the primary, and early contractions during labor occur here

_____11. location of placenta
_____12. location of fetus during the third trimester
_____13. the zygote is formed here

_____14. muscle tissue

_____15. nervous system

_____16. skin

_____17. bone

_____18. ovaries

Use the following answers to give the germ layer from which the items on the left were derived.
a. ectoderm
b. endoderm
c. mesoderm

Multiple-Choice Questions

Circle the choice that is the best answer for each question.

19. In frogs, the central nervous system is one of the
    a. systems which is already present in the gametes.
    b. first systems to develop.
    c. systems which appears about midway in development.
    d. last systems to develop.

20. Placental mammals probably have eggs with a(n) __________ amount of yolk.

a. essentially nonexistent
b. small
c. medium
d. large

21. The gender of a human is determined
   a. during the first trimester.
   b. during the second trimester.
   c. during the third trimester.
   d. after birth.
   e. at puberty.

22. In chicken embryos, the _________ is the site for exchanging carbon dioxide with the environment.
   a. yolk
   b. allantois
   c. heart
   d. archenteron membrane
   e. chorioallantoic membrane

Fill in the Blanks

Complete each statement by writing the correct word or words in every blank.

23. The time between conception and birth in humans is approximately _________ days.

24. The ultimate specialization of a cell is determined by the cell's _________.

25. The three _________ layers that appear in the embryo are _________ on the outside, _________ as a middle layer, and _________ as the innermost layer.

26. In humans, cell division after fertilization is called _________. This begins in the _________ of the woman.

27. Most human cells become specialized during the _________ trimester.

Questions For Further Thought

Write your answer for each question in the space provided.

28. Adult humans are not able to regenerate lost arms. What information must scientists come up with before the body can be made to show this type of repair?

SUMMARY

Study the seventeen items listed in the SUMMARY on p. 470 of the text. It should be clear that embryonic development occurs in an orderly sequence in animals. Although the controls and mechanisms are not always understood by biologists, it *is* possible to predict which groups of cells will develop into specific adult body parts at very early stages of development. Ultimate control, as you remember, lies in turning genes on or off in the cells. Developing offspring can be very delicate and may easily be harmed by outside factors, especially early in their growth.

HINTS

Read these suggestions before you check your answers to the EXERCISES.

For question 4: They are all vertebrates.

For question 6: The placenta is also called the afterbirth.

For question 8: Remember what you learned in Chapter Fifteen.

For question 12: By that time it is fairly large.

For question 20: The answer is not "A."

For question 21: This asks when is it set, not when do the parents figure it out.

For question 24: Remember what Mendel and others said.

For question 26: Don't use mitosis for the first answer.

CONGRATULATIONS . . . YOU HAVE COMPLETED CHAPTER 16 !!!

Part Five Systems and their Control

CHAPTER 17

SUPPORT AND MUSCULAR SYSTEMS

CONNECTIONS

You are beginning Part Five of the text, focuses primarily on humans, but also examines the body systems of other representative animal phyla. As you study these chapters, look for relationships between how a body is built and what it is able to do.

Your bones and muscles are obvious body parts that you are aware of all of the time. Now you have an opportunity to learn about them in detail. This chapter shows you the many functions and interactions of various types of skeletal and muscular systems in the animal kingdom. Watch for the common features of these systems as well as for the unique aspects of each of the different ones. There is not much chemistry in this chapter, but be sure to pay attention to the nature and actions of the different cell types of each system.

OBJECTIVES FOR THIS CHAPTER

When you have finished Chapter 17, you should be able to:

1. List the functions of both skeletal and muscular systems.
2. Use examples and describe the various types of skeletal support systems.
3. Describe the functions and physical properties of cartilage.
4. Describe the functions and physical properties of bone.
5. Differentiate between the appendicular and axial skeleton.
6. Explain the structure and action of the four types of joints.
7. Describe the contractile systems of a cnidarian, a roundworm, and an arthropod.

8. Describe the structure and function of the three types of vertebrate muscles.
9. Explain what is meant by antagonistic muscle pairs and describe how these work to move bones.
10. List the components of a skeletal muscle fiber and describe how each of the components functions during muscle contraction.

CHAPTER 17 OUTLINE

I. Support Systems
   A. Types of Skeletal Systems
   B. Types of Connective Tissue
   C. The Human Skeleton
   D. Joints
II. The Variety of Contractile Systems
   A. Types of Muscles in Humans
   B. Skeletal Muscles
      1. Antagonistic Muscles
   C. Muscle Contraction

CONCEPTS IN BRIEF

Support Systems

Although most skeletal systems contain hard parts, the hydrostatic skeletons of cnidarians, flatworms, and annelids function by compressing fluids, which are in the body cavity. The resulting pressure causes the animal to move. Remember that any skeletal system provides support and assists in movement, and these functions can be accomplished in many ways. The skeletal system of a sponge gives the animal body its form.

An exoskeleton provides a certain amount of protection, but it may also present unique problems as the animal grows. Common solutions to this problem include adding on to a shell, or throwing away the old shell and producing a new one.

The endoskeletons of vertebrates all contain cartilage, and most also have bone. These two types of connective tissue (along with tendons and ligaments) are emphasized in the chapter. Bone is active living tissue that is primarily composed of cells, collagen matrix, marrow, and salts of calcium and phosphorus.

Cartilage cells and matrix make up the cartilage tissue, which is more flexible than bone and is also smooth. You can feel your own cartilage in your ears and the tip of your nose. In addition it is found in many moveable joints. Tendons and ligaments provide additional support and attach parts of the muscular and skeletal systems to each other.

Your entire skeleton can be divided into the axial and appendicular skeletons, which are composed of all of your bones (approximately 206 bones in most people). Bones move (relative to each other) at joints, but at some joints the bones are fused and movement is no longer possible. It is easy for you to watch a moving body and determine the amount and direction of movement that a joint is able to show.

The Variety of Contractile Systems

Cnidarians use contractile cells in order to move, but since they do not have mesoderm these contractile cells do not make up true muscles. Beginning with the roundworm phylum, mesoderm is present as a cell type, and the term "muscle" is appropriate to use. Keep in mind that these cells produce movement by shortening, thereby causing regions of the body to move closer to each other. When these cells relax, and return to their previous length, other cells must shorten in order to cause the body parts to move back where they were.

An Arthropod's muscles are on the inside of its exoskeleton. This arrangement allows them to have rapid wing beats and great strength for their size. One reason that people are soft to the touch is due to the fact that our skeletal muscles (the ones that make your bones move) are on the outside of our endoskeleton. These are also called the voluntary muscles, since they are the ones over which you have the most conscious control.

Many skeletal muscles are arranged in pairs, in which the two muscles work in opposition to one another. This antagonistic arrangement allows you to have fine control over your body, whether it is for balance or for movement in any of a variety of different directions.

The physiology of muscle contraction involves energy exchange and the rapid movement of filaments. When calcium is present, the myofilaments called myosin attach their cross bridges to the myofilaments called actin. The actin molecules are then pulled toward one another. Since myofibrils are made of these myofilaments, this causes the myofibrils to shorten. Now, since these myofibrils are what makes up a muscle fiber, the muscle fiber (cell) shortens. Finally, since what you call a muscle is really very many muscle fibers held together by connective tissue, the muscle also shortens (contracts) and your bones move. As you would have guessed, ATP supplies the energy.

KEY TERMS

Be sure to make and use flash cards for all of these terms. Suggestions are found in "To The Student" at the beginning of the Study Guide. Page numbers refer to the text.

actin	493	ligament	481
antagonist muscle	489	marrow	479
appendicular skeleton	484	matrix	479
axial skeleton	482	muscle fiber	488
ball-and-socket joint	486	myofibril	492
bone	479	myofilament	493
canaliculi	479	myosin	493
cardiac muscle	488	organs	477
cartilage	481	organ systems	477
collagen	479	origin	489
connective tissue	479	skeletal muscle	488
cross bridge	493	skeletal system	478
endoskeleton	478	skull	482
exoskeleton	478	sliding filament theory	493
extensor	489	smooth muscle	488
fascia	489	spicule	478
flexor	489	spinal column	482
Haversian canal	479	spongin	478
Haversian system	479	striation	488
hinge joint	486	suture line	486
hydrostatic skeleton	478	symphysis	486
insertion	489	tendon	481
joint	485	tissues	477

EXERCISES

Check your understanding of Chapter 17 by completing the following exercises. Answers begin on p. 207.

1. Use the following drawings to answer the questions below.

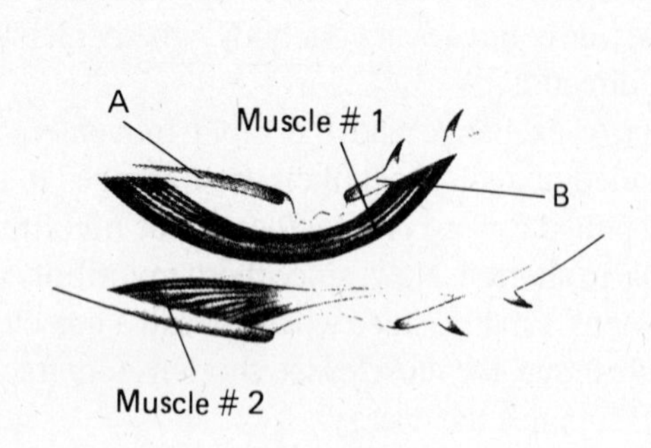

Insect

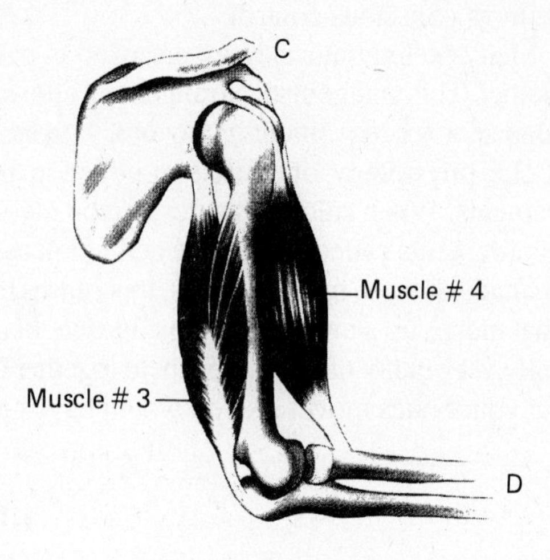

Human

a. In order for the insect to bring point A closer to point B, which muscle would have to contract?

b. In order for the insect's limb to go back to its original position, what must occur?

c. In order for the human to bring point C closer to point D, which muscle would have to contract?

d. In order for the human's arm to go back to its original position, what must occur?

True or False Questions

Mark each question either T (True) or F (False).

____2. In order to pull something toward you, your muscles must shorten, but for the force needed to push something away your muscles must cause themselves to get longer.

____3. Muscle cells have many mitochondria.

____4. Humans have true muscle.

____5. The metabolic activity in your bones stops when you have grown to your full height.

____6. Myofilaments are made of muscle cells.

____7. Cartilage is more flexible than bone.

____8. Cardiac muscle is a type of involuntary muscle.

Matching Questions

Each question may have more than one answer; write all of the answers that apply. Answers may be used more than once.

____9. present in insects

____10. skeletal muscles are inside the skeleton

____11. may be shed and replaced as the animal grows

a. endoskeleton
b. exoskeleton
c. both a and b
d. none of the above

Use the letters on the diagram to the right as the possible answers for Questions #12–19.

____12. a hinge joint

____13. a ball and socket joint

____14. an immobile joint

____15. part of the pelvic girdle

____16. part of the pectoral girdle

____17. the humerus

____18. the radius

____19. cartilage is found here

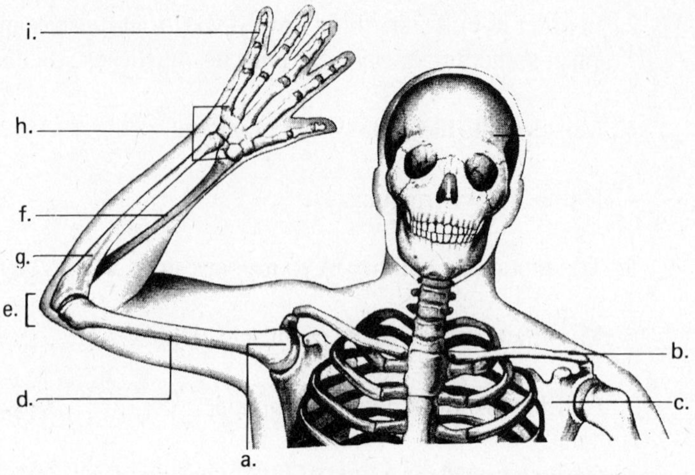

Multiple-Choice Questions

Circle the choice that is the best answer for each question.

20. A group of cells that are clustered together and function collectively to perform a particular task would be a(n)
    a. amoeba.
    b. tissue.
    c. organ.
    d. organ system.

21. Which of the following types of joints allows the greatest freedom of movement?
    a. ball-and -socket
    b. hinge
    c. symphysis
    d. suture line

22. Earthworms have a(n)
    a. axial skeleton.
    b. endoskeleton.
    c. exoskeleton.
    d. hydrostatic skeleton.

23. According to the sliding filament theory, the _____ molecule causes _____ molecules to move.
    a. actin, myosin
    b. actin, calcium
    c. myosin, actin
    d. myosin, calcium

Fill in the Blanks

Complete each statement by writing the correct word or words in every blank.

24. The mitochondria in a muscle cell provide the cell with _________.

25. Your femur is held to your fibula by _________.

26. The blood vessels which serve bone cells run through an opening in the bone called the _________ _________.

27. Your skull is part of your _________ skeleton.

28. Your skull contains _________ joints.

Questions For Further Thought

Write your answer for each question in the space provided.

29. Explain why it is important to continue to take in calcium even though you may no longer be growing in height.

SUMMARY

Study the fifteen items listed in the SUMMARY on p. 493 of the text. You have seen that organisms gain shape, protection, support, and the ability to move as a result of the nature and interaction of their muscular and skeletal systems. Although the emphasis in this chapter has been on human systems, there are a variety of ways in which animals have accomplished these results.

Once again it is clear that the activities of the cells are responsible for the activities of the organism. This chapter has also introduced the idea that organized groups of cells, or tissues, can function together in a specialized way within the body.

HINTS

Read these suggestions before you check your answers to the EXERCISES.

For question 1:	These are antagonistic pairs. Also notice that the insect muscles are inside the skeleton and the human muscles are outside the skeleton.
For question 5:	Bone is live tissue.
For question 6:	Which one of these is larger?
For question 19:	Cartilage is found in joints.
For question 25:	The answer is not "muscles."
For question 29 :	Remember to consider bones *and* muscles.

CONGRATULATIONS . . . YOU HAVE COMPLETED CHAPTER 17 !!!

CHAPTER 18

HOMEOSTASIS AND THE INTERNAL ENVIRONMENT

CONNECTIONS

When you studied cells you learned that they are able to control the passage of many substances which can move through their membranes. By doing this they maintain unequal concentrations of materials on the inside and outside of their plasma membranes.

In this chapter you will apply your knowledge of cells (and a bit of chemistry) to the ways in which animals attempt to maintain a relatively constant set of conditions in their bodies. Be sure to notice the effect of the environment on the organism and its responses. The emphasis here is on temperature controls and water/waste balance in vertebrates, but other organisms are also considered. Watch for common features in the strategies which are used to solve these problems.

OBJECTIVES FOR THIS CHAPTER

When you have finished Chapter 18, you should be able to:

1. Define homeostasis.
2. Compare and contrast positive and negative feedback systems.
3. Differentiate between endotherms and ectotherms and explain how each regulates temperature to maintain homeostasis.
4. Name three main nitrogenous waste products and describe their levels of toxicity.
5. Compare and contrast the relationship of waste removal and maintenance of water balance in freshwater protists and freshwater fish.
6. Compare and contrast the mechanisms by which freshwater, marine, and terrestrial vertebrates eliminate their nitrogenous waste and regulate water balance.
7. Name the structures in the human excretory system and give the functions of each.
8. Describe the formation of urine in the human kidney.
9. Describe the role of hormones in controlling the concentration of urine.

CHAPTER 18 OUTLINE

I. Homeostasis and the Delicate Balance of Life
II. Feedback Systems
III. Homeostasis and the Regulation of Temperature
  A. Ectotherms
  B. Endotherms
IV. Homeostasis and the Regulation of Water
  A. Solutions to the Water Problem
  B. Invertebrate Excretory Systems
  C. The Human Excretory System
    1. The Formation of Urine
    2. The Concentration of Urine

CONCEPTS IN BRIEF

Homeostasis and the Delicate Balance of Life

You know that responding to the environment is one of the characteristics of a living thing. Homeostasis involves responses which allow the organism to keep the conditions of its internal environment fairly constant. Homeostasis does not mean staying exactly the same.

Feedback Systems

In order to maintain a steady state, most organisms use negative feedback systems rather than positive feedback systems. The feedback aspect of this relies on the body monitoring its own conditions and communicating with itself about what is going on. Negative feedback results in increasing whatever there is too little of, or decreasing something if there is too much.

Homeostasis and the Regulation of Temperature

Each kind of living thing can only survive in a certain range of temperatures. If it gets too cold or hot the organism will die, due to the formation of ice crystals or the disruption of its body chemistry. Ectotherms produce body heat (remember the laws of thermodynamics?) but they aren't as efficient at keeping it or regulating its production as are the endotherms. Consequently ectotherms regulate their body temperature by changing their behavior.

Endotherms may also change behavior in order to regulate temperature, but they generally rely on their physiology to heat or cool themselves. This may require a large energy expenditure, however it allows the animal to maintain its level of activity over a wider range of environmental conditions.

Human temperature regulation involves changes in blood flow, hormone levels, perspiration, metabolic rate, muscle activity, and behavior. Fever is a normal response (within limits) that raises the body temperature as a means of fighting infection. High levels of water in a body give it a certain amount of temperature stability, since water (as you remember from Chapter Three) changes temperature slowly.

Homeostasis and the Regulation of Water

Animals have combined the processes of water regulation and the removal of toxic nitrogenous wastes. Different species lose water by using high, medium, or low amounts of it to dilute these nitrogenous wastes (the result of protein digestion), which are excreted as either ammonia, urea, or uric acid accordingly. In addition to waste removal, saltwater fishes also lose water to the environment by osmosis.

Lost water must be replaced by drinking, acquiring it from the breakdown of food (remember the end of the electron transport chain?), or having it move into cells by osmosis from the environment.

The nephridia in the segments of an earthworm operate on a plan that is similar to that of the nephrons in a human kidney. That is: fluids move into a tube, desirable materials are reabsorbed from the tube into the blood, and whatever is left in the tube is excreted to the environment as waste.

Osmotic concentrations regulate the movement of materials in nephrons of the kidney. It may seem like there is a lot of movement of the same materials in and out of the different regions of a nephron; and there is! It is the combination of filtration, reabsorption, and secretion that produces the final result: keeping the desired chemicals and throwing away the wastes. Additional control of urine formation is achieved by the interaction of hormones which, of course, are affected by what an animal eats or drinks.

KEY TERMS

Be sure to make and use flash cards for all of these terms. Suggestions are found in "To The Student" at the beginning of the Study Guide. Page numbers refer to the text.

Term	Page	Term	Page
aldosterone	512	loop of Henle	508
ammonia	504	negative feedback	498
antidiuretic hormone (ADH)	512	nephridia	508
atrial natriuretic		nephron	508
factor (ANF)	512	positive feedback	499
Bowman's capsule	508	proximal convoluted tubule	508
collecting duct	509	pyrogen	503
distal convoluted tubule	508	reabsorption	510
ectotherms	500	secretion	510
endotherms	501	urea	504
estivate	501	ureter	509
filtration	510	urethra	509
glomerulus	505	uric acid	505
homeostasis	498	urinary bladder	509
hypothalamus	502	urine	504

EXERCISES

Check your understanding of Chapter 18 by completing the following exercises. Answers begin on p. 207.

Use the following drawings to answer Questions #1–6. Write the correct letter(s) of the animal(s)in the blanks. Each question may have more than one answer.

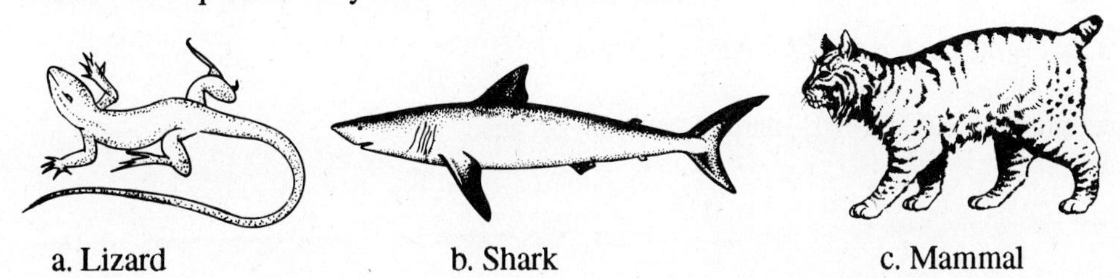

a. Lizard b. Shark c. Mammal

____1. Endotherm

____2. Ectotherm

____3. Regulates body temperature by moving.

____4. Lives in an environment with a stable temperature.

____5. Normally uses negative feedback mechanisms to regulate its body temperature.

____6. Activity level slows in colder conditions.

True or False Questions

Mark each question either T (True) or F (False).

_____7. Homeostasis doesn't involve keeping conditions exactly the same. It really involves keeping conditions within a range.

_____8. Ectotherms are cold blooded because they do not produce any body heat.

_____9. Fish that live in shallow water have a more stable environment than fish which live in deep water.

_____10. Raising your body temperature by exercising may help to prevent certain infections.

Matching Questions

Each question may have more than one answer; write all of the answers that apply. Answers may be used more than once.

_____11. what water moves into when it first enters the nephron from the blood

_____12. where urea moves out of the tubule systems and into the medulla

_____13. major site(s) of water reabsorption

_____14. collects and holds urine from both kidneys

a. collecting duct
b. urinary bladder
c. Bowman's capsule
d. loop of Henle
e. glomerulus
f. capillaries around the descending loop

Multiple-Choice Questions

Circle the choice that is the best answer for each question.

15. If their temperature increases slowly, animals will die because
    a. they cook.

b. ADH concentrations are lowered.
c. hydrogen bonds of their proteins break.
d. the reabsorption of aldosterone increases too rapidly.

16. A freshwater fish will
    a. not release any nitrogenous waste.
    b. have watery urine.
    c. lose water from its cells to the environment by osmosis.
    d. all of the above.

17. Birds secrete uric acid, which requires them to also secrete _____ water.
    a. very little
    b. a moderate amount of
    c. large amounts of

18. A diet high in _____ would be best for water regulation in animals that cannot replace their water easily.
    a. fat
    b. salt
    c. urea
    d. protein

19. The action of ADH is _____ by ethyl alcohol, therefore you will need to drink _____ water.
    a. suppressed, less
    b. suppressed, more
    c. enhanced, less
    d. enhanced, more

Fill in the Blanks

Complete each statement by writing the correct word or words in every blank.

20. An infection can cause the body to set its __________ higher, and the result is a __________. This may kill temperature-sensitive __________ and __________.

21. Wastes containing __________ are formed from protein digestion.

22. The processes of water regulation and removal of metabolic wastes rely on separate mechanisms in freshwater __________, but they use the same mechanisms in __________.

23. Urine formation in your kidneys relies on the three processes of __________, __________, and __________.

Questions For Further Thought

Write your answer for each question in the space provided.

24. Explain why placing a thermometer under someone's tongue gives a more accurate body temperature than having them hold it in their fist.

25. Why is dry dirt harmful to worms?

SUMMARY

Study the ten items listed in the SUMMARY on p. 513 of the text. You have seen that an organism's internal and external environment may change minute by minute. In order to survive, the organism must be able to either move, tolerate these changes, or alter itself to accommodate the environment. This chapter has shown that there are a variety of ways in which organisms respond to changes in the temperature and osmotic balance of their surroundings, while they throw away toxic nitrogenous wastes. This adjusting and readjusting is a constant activity for organisms.

HINTS

Read these suggestions before you check your answers to the EXERCISES.

For question 3:	Moving may not be the *only* method it uses.
For question 5:	Remember, positive feedback systems are not common in animals.
For question 6:	Assume that this mammal does not hibernate.
For question 8:	Don't forget the second law of thermodynamics.
For question 11:	The key word here is "first."
For question 15:	"A" is *not* the answer.
For question 20:	The second answer is "fever."
For question 22:	The first answer is *not* "fish."

CONGRATULATIONS . . . YOU HAVE COMPLETED CHAPTER 18 !!!

CHAPTER 19

THE RESPIRATORY, CIRCULATORY, AND DIGESTIVE SYSTEMS

CONNECTIONS

All living things are subject to the effects of the second law of thermodynamics. Therefore, since they must replace their energy supplies on a regular basis, a discussion of digestion is in order. This chapter considers aerobic organisms, so it is also appropriate to learn how organisms obtain and distribute the oxygen which they will use during the production of ATP. Finally the circulatory system must also be considered in order to see how the digested food and the oxygen get to the cells, and how the wastes are removed.

As always, look for the common ways in which different organisms solve the same problems. The examples used in this chapter illustrate that it is also possible to see many variations of common solutions. Think about the effects of natural selection on body systems as you study this material.

OBJECTIVES FOR THIS CHAPTER

When you have finished Chapter 19, you should be able to:

1. Compare the mechanisms by which aquatic and terrestrial animals obtain oxygen from the environment.
2. Compare the mechanisms by which aquatic and terrestrial animals deliver oxygen to the cells of their bodies.
3. Describe the components of the mammalian respiratory system, and explain the function of each.
4. Explain how humans ventilate their lungs.
5. Compare the circulatory systems of small and large animals and explain why they are different.
6. Discuss the structural variation in representative vertebrate hearts and describe how blood moves through each.
7. Explain what causes the vertebrate heart to beat.
8. Compare the mechanisms used for digesting food by representative invertebrates and vertebrates.
9. Trace the path of nutrients through the various parts of a typical vertebrate digestive system.

10. List the structures in the human digestive system and describe how each functions during digestion.

CHAPTER 19 OUTLINE

I. Respiratory Systems
  A. The Various Ways Species Get Oxygen
    1. Aquatic Animals
    2. Insects
    3. Mammals
II. Circulatory Systems
  A. Circulation in Small Animals
  B. Circulation in Larger Animals
    1. The Vertebrate Vessels
    2. Arteries and Blood Pressure
    3. Blood
    4. The Vertebrate Heart
    5. Control of Heartbeat
  C. The Lymphatic System
III. Digestive Systems
  A. Digestive Arrangements
  B. The Human Digestive System

CONCEPTS IN BRIEF

Respiratory Systems

Oxygen moves across *moist* membranes when it enters cells, so as you examine the various types of respiratory systems in this chapter be sure to notice that the exchange surfaces are either in water or are kept wet in some fashion. In addition, the exchange surface must be large enough to diffuse the amount of oxygen required by all of the cells of the organism (remember the surface:volume ratio from Chapter Four).

Smaller animals may use their entire body surface as a gas exchange area, but larger animals must have a specialized body region for exchange plus a method of moving the gasses to and from all of its cells. Vertebrates rely on numerous thin layers of cells (which have rich blood supplies) in gills and lungs for obtaining oxygen and releasing carbon dioxide. Aquatic vertebrates must keep a flow of water passing over their gills, and mammals constantly inhale and exhale to keep a flow of air moving through their lungs. Humans rely on carbon dioxide sensors and involuntary responses to regulate their rate of breathing.

Terrestrial insects have a unique tracheal system which brings air into the body segments through numerous openings and distributes it directly throughout the body. Their respiratory system does not rely on their circulatory system for delivery of oxygen and carbon dioxide.

Circulatory Systems

All organisms must move gasses, nutrients, wastes, and miscellaneous molecules within their bodies. Small organisms may be able to use **diffusion** as their sole means of doing this if all of the cells are close to an exchange surface (or are even in direct contact with the environment). Larger animals, however, require a circulatory system of some sort.

Saclike structures which extend throughout the body (and double as digestive systems) are used for circulation by cnidarians and flatworms. Larger animals use some sort of vessels which contain fluids. Some mollusks and arthropods have an **open circulatory system** in which hemolymph (their blood) is pumped through vessels but then flows out of the vessels and bathes the tissues directly, as it slowly moves back to the heart.

Closed circulatory systems with hearts (in which the blood stays in the vessels) are used by annelids and vertebrates. This provides an efficient way to move materials to and from regions of the body that are not close to any exchange surfaces. The vessels branch into many very small, thin capillaries which are able to get near every cell in the body. Exchange of materials with cells takes place through the capillary walls.

Blood is composed of plasma (a liquid with many chemicals in solution) and various types of cells. Since hemoglobin in the red blood cells is able to combine with oxygen, human blood can carry larger amounts of oxygen than it could without hemoglobin. Small amounts of carbon dioxide are carried in solution and by hemoglobin, but most of it circulates as bicarbonate ions.

Vertebrates use a **heart** to pump blood through their circulatory system. The chamber which pumps blood away from the heart is an atrium. A ventricle of the heart receives blood from the body and pumps it to an atrium. The vertebrate hearts with more chambers and better separation of oxygen-rich and oxygen-poor blood provide more efficient blood pressure and oxygen delivery throughout the body of the animal. Heart muscle beats automatically, but the rate is regulated by nodal tissue and the impulse is spread to the ventricles by Purkinje fibers.

Some plasma leaves the capillaries and flows over the body cells of vertebrates. These fluids are collected by the vessels of the **lymphatic system**, which conduct them back to veins near the heart. Distributed along the vessels are nodes, in which white blood cells remove cellular debris and bacteria from the fluids.

Digestive Systems

Large molecules are broken down into smaller ones during digestion. These smaller molecules are then moved throughout the organism's body by the circulatory system (or by diffusion in organisms without a specialized circulatory system). Cnidarians and flatworms have a digestive system with a single opening which must take in food and also expel wastes. Digestion of food particles begins in the gastrovascular cavity (outside of the cells) and is completed inside of specialized cells.

Annelids and vertebrates are examples of organisms with a two-opening system in which digestion takes place outside of the cells and the digested materials are then absorbed. This is the "tube within a tube" body plan (from Chapter Sixteen) with specialized regions for: breakdown of food, absorption, and collection and removal of wastes.

Humans use mechanical breakdown to grind food into smaller pieces in the mouth and stomach. This is accompanied by extensive chemical activity (using enzymes) to complete digestion in the mouth, stomach, and small intestine. The efficiency of absorption (in the small intestine) is increased by numerous villi and microvilli which increase the overall surface (absorptive) area.

The accessory organs (pancreas, liver, and gall bladder) add secretions to the small intestine which aid in digestion. Undigested materials and waste products collect in the large intestine where water and vitamins (secreted by *E. coli*) are absorbed into the body cells. All remaining materials (feces) are eliminated through the anus.

KEY TERMS

Be sure to make and use flash cards for all of these terms. Suggestions are found in "To The Student" at the beginning of the Study Guide. Page numbers refer to the text.

air sacs	521
alveoli	521
aorta	529 & 531
aortic arches	526
arteriole	527
artery	526
atrioventricular (AV) node	534
atrium	531
bile	543
breathing	521
bronchi	521
bronchiole	521
capillary	527
carbonic anhydrase	531
cellular respiration	517
closed circulatory system	526
crop	539
diastole	529
diastolic pressure	529
digestion	538
erythrocytes	530
esophagus	540
external respiration	517
feces	545
food vacuole	538
gizzard	539
gall bladder	543
heart	531
hemoglobin	530
hemolymph	526
hepatic portal vein	543
internal respiration	517
lacteal	542
villi	542
large intestine	545
leukocytes	530
liver	543
lung	521
lymph	536
lymph capillaries	537
lymph duct	538
lymph node	537
lymphatic system	536
lymphocyte	538
microvilli	542
nodal tissue	534
open circulatory system	526
pancreas	542
peristalsis	540
plasma	529 & 537
platelets	530
Purkinje fibers	536
rectum	545
respiration	517
sinoatrial (SA) node	534
sinus	526
small intestine	542
spiracle	520
stomach	540
systole	529
systolic pressure	529
trachae	521
tracheal system	520
vein	527
ventilation	521
ventricle	531
venule	527

EXERCISES

Check your understanding of Chapter 19 by completing the following exercises. Answers begin on p. 207.

Use the following drawings to answer Questions #1–4. Each question may have more than one answer.

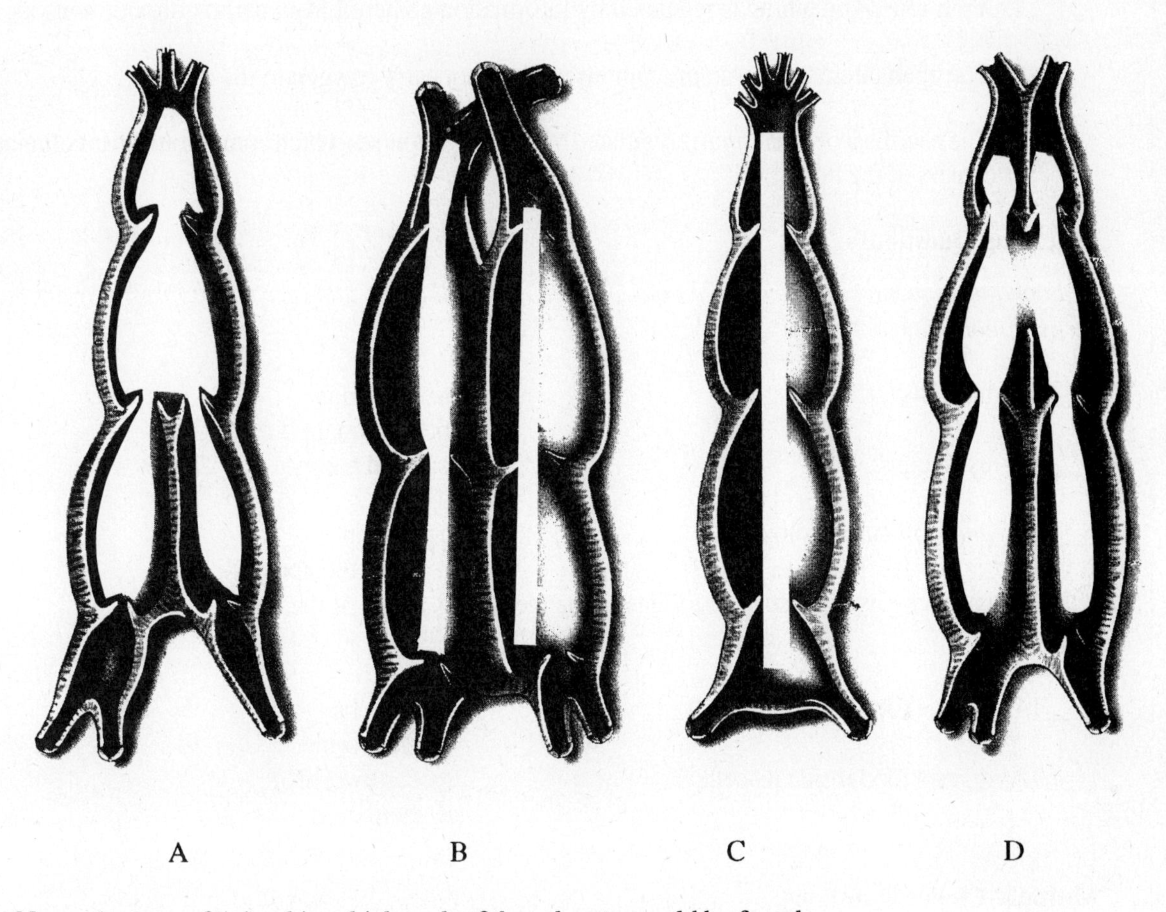

1. Name the type of animal in which each of these hearts would be found.

   a. ____________________ b. ____________________

   c. ____________________ d. ____________________

2. Add arrows to show the direction of blood flow through each heart.

3. In which of these hearts would you find oxygen-rich and oxygen-poor blood mixed together?

4. In which of these animals does the blood return to the heart after it passes through a capillary bed, but before it passes through another capillary bed?

True or False Questions

Mark each question either T (True) or F (False).

_____5. Gas exchange surfaces must be moist in order to function well.

_____6. Birds have a single-opening digestive system.

_____7. Your rate of breathing is regulated by information gathered from carbon dioxide sensors.

_____8. The open circulatory system of an insect does not carry oxygen to the animal's cells.

_____9. The heartbeat of a mammal is regulated by nervous impulses which come from the involuntary centers of the brain.

Matching Questions

Each question may have more than one answer; write all of the answers that apply. Answers may be used more than once.

____10. has gills

____11. uses lungs

____12. gives off carbon dioxide

____13. uses only its body surface for gas exchange

a. cnidarian
b. mammal
c. mud puppy
d. fish
e. human
f. all of the above
g. none of the above

____14. digests food outside its cells

____15. digests food inside its cells

a. amoeba
b. flatworm
c. earthworm
d. human

Multiple-Choice Questions

Circle the choice that is the best answer for each question.

16. Carbon dioxide will move out of the body of a fish if the level of_____ is _____ in its body than in the water.
   a. oxygen, higher
   b. water, higher
   c. carbon dioxide, higher
   d. none of the above

17. Humans take oxygen from the air, and release carbon dioxide into the air in the
   a. nose.
   b. pharynx.
   c. bronchioles.
   d. alveoli.
   e. all of the above

18. Nutrients and wastes move in and out of your circulatory system in your
    a. arteries.
    b. veins.
    c. capillaries.
    d. all of the above
    e. none of the above

19. The large intestine
    a. collects wastes.
    b. absorbs water.
    c. absorbs certain vitamins.
    d. contains bacteria on the inside.
    e. all of the above

Fill in the Blanks

Complete each statement by writing the correct word or words in every blank.

20. Oxygen moves from the blood to the body cells by the process of __________.

21. In order to avoid lung problems in the future, smokers should __________ smoking.

22. The muscles which are used by mammals for inhaling and exhaling are the __________, and the muscles between the __________.

23. During a heartbeat in a mammal, the two __________ contract, and then the __________ contract.

24. The products of fat digestion enter the body through the __________ of the many __________ in the small intestine.

Questions For Further Thought

Write your answer for each question in the space provided.

25. Describe how the surface area of your small intestine has been increased without making the intestine longer. Why is this beneficial?

SUMMARY

Study the nineteen items listed in the SUMMARY on p. 546 of the text. All of the organisms mentioned in this chapter use oxygen, break down food, and move materials within their bodies. You have seen that actions of the respiratory, circulatory, and digestive systems of many organisms are closely linked together, while in other organisms they are entirely separate from one another. These relationships are the result of natural selection and the influence of the environment on the organism.

It should also be clear that each of the various styles of building and operating a system has its own advantages and disadvantages. This, of course, places limits on what any organism can do. Limits, however, have not kept organisms from being successful in their habitats, they simply give the organisms their unique characteristics.

HINTS

Read these suggestions before you check your answers to the EXERCISES.

For question 3:	Think hard about the pattern of blood flow in a fish.
For question 4:	Think about fish again.
For question 9:	Don't forget about the SA node.
For question 14:	Do some of these organisms use both?
For question 16:	Remember the rules of diffusion.
For question 17:	Think about the requirements for a gas exchange surface.
For question 24:	The first answer is not "microvilli," nor is it "villi."

CONGRATULATIONS . . . YOU HAVE COMPLETED CHAPTER 19 !!!

CHAPTER 20

THE IMMUNE SYSTEM

CONNECTIONS

With literally millions of different kinds of living things on the earth it isn't surprising that quite a few make their living at the expense of your body. You constantly have invaders (especially viruses, bacteria, protists and fungi) in and on yourself. Without protection your cells would soon be overwhelmed, and since you *are* your cells, you would become ill (and eventually dead).

This chapter describes the various ways in which the immune system works to keep your body healthy, and to promote its recovery when you are temporarily losing the battle. You should notice that there are many different responses to invaders in both a specific and a general sense. Don't get lost in the details and lose sight of the overall plan of protecting the cells.

OBJECTIVES FOR THIS CHAPTER

When you have finished Chapter 20, you should be able to:

1. Distinguish between specific and nonspecific immune responses.
2. Name the nonspecific immune responses and state how each defends the body against infection.
3. List five major types of white blood cells and describe their roles in human defense responses.
4. Describe the structure and function of an antibody.
5. Describe how each type of B and T lymphocyte functions during specific immune responses.
6. Discuss how vaccinations result in immunity to diseases.
7. Explain what happens during autoimmune reactions.
8. Explain the role of interferon in protection against viruses.
9. Describe the acquired immune deficiency syndrome, its symptoms, related diseases, and modes of transmission.
10. Discuss the possible connection between one's state of mind and physical well being.

CHAPTER 20 OUTLINE

I. The Nonspecific Responses
II. Specific Responses
  A. The Antigen-Antibody Response
  B. Programming the Lymphocytes
  C. The Role of the Cytotoxic T-Cells
  D. The Roles of the Helper T-Cells and the B-Cells
  E. Tolerance and Autoimmunity
III. Interferon: An Exciting New Promise
IV. AIDS: A Devastating New Problem
V. Mind and Body

CONCEPTS IN BRIEF

The Nonspecific Responses

Preventing all invaders from entering the body is your first line of defense, and is part of the nonspecific response system. Remember that you are a tube within a tube, so both the lining of your digestive tract and your skin are "outer" coverings. Table 20.1 lists the components of these barriers and their basic protective functions. The combined action of all of these mechanisms provides a very successful barrier system.

Initiating an inflammatory response occurs when an invader succeeds in breaking through the defenses described above. The soreness, swelling, redness, and fever of nonspecific responses are probably familiar to you. Although you may have always treated these signs as a bother, each one of these acts to prevent the spread of an invader.

Specific Responses

In addition to the nonspecific responses above, your immune system also initiates specific responses. These target each different type of invader with a variety of short and long term defenses to stop an infection and to prevent future ones.

The actions of white blood cells (macrophages, and the lymphocytes known as B-cells and T-cells) are important here. Each type of invader has a unique antigen structure which identifies it. These antigens trigger two responses:

(1) macrophages attack and dismantle any invaders which they find, and the accompanying antigens then appear on the outside of the macrophage

(2) B-cells attach to the antigens of invaders that have not been consumed by macrophages and hold on to them.

T-cells now are able to attach to the foreign antigens on the macrophages.

If interleukin is present, it stimulates cell division in both the T-cells and B-cells mentioned in items 1 and 2 above. The T-cells produce cytotoxic T-cells and cytotoxic memory cells. The B-cells produce memory B-cells and plasma cells. All of these new cells have specific functions (see Table 20.2). The combination of their actions produces effective primary and secondary responses to each particular type of invader.

Primary responses may require several days before they are effective in controlling the infection. Secondary responses occur very rapidly if there is a future infection, which is the concept on which vaccinations have been developed.

Generally your immune system recognizes the difference between your normal body cells and invaders, and only attacks invaders. When body cells *are* attacked (an autoimmune reaction) severe problems may result, including one form of arthritis.

Interferon: An Exciting New Promise

Interferon is a natural protein which your cells produce in order to prevent viruses from attacking healthy cells. Its generalized effects on any viral infection have resulted in a great deal of research in hopes of producing effective medical treatments against viruses. Various tests have shown mixed results, and it is hoped that genetic engineering will increase our understanding of possible uses.

AIDS: A Devastating New Problem

Acquired immune deficiency syndrome (AIDS) is caused by the human immunodeficiency virus (HIV). This virus attacks T-cells and macrophages, and the result is an ineffective immune system. Consequently the patient becomes susceptible to many infections that he or she would normally have controlled with the immune responses described above. At the present time AIDS is fatal, and it takes an average of ten years (after infection) to show symptoms.

HIV is transmitted through body fluids, with common methods including: intercourse, transfusions using infected blood, sharing needles by intravenous drug users, and infected mothers passing it to either the developing fetus or the baby during birth. Effective preventative measures include barriers (rubber gloves and condoms), washing your hands (it is a fragile virus), and handling only sterile needles (or other medical instruments).

Mind and Body

Although our understanding of the mechanism is not complete, it is clear that your mental state can influence your health. In particular, stress which is not managed can suppress the immune system and lower white blood cell counts, resulting in illness. The effects of neuropeptides are being studied as a possible link between brain activity and the immune system.

KEY TERMS

Be sure to make and use flash cards for all of these terms. Suggestions are found in "To The Student" at the beginning of the Study Guide. Page numbers refer to the text.

Term	Page
acquired immune deficiency syndrome (AIDS)	564
antibody	557
antigen	557
antigen recognition site	558
autoimmunity	562
basophil	553
B-cells	557
cellular immunity	557
constant region	557
cytotoxic T-cells	557
eosinophil	553
heavy chains	557
histamine	553
human immunodeficiency virus (HIV)	564
humoral immunity	557
immune system	551
inflammatory response	553
interferon	562
interleukin II	558
keratin	551
light chains	557
lymphocyte	553
lysozyme	552
macrophage	554
memory B-cell	559
monocyte	553

EXERCISES

Check your understanding of Chapter 20 by completing the following exercises. Answers begin on p. 207.

1. Arrange the following events of the specific immune response in sequence, from first to last.
   a. interleukin stimulates B-cells to divide
   b. foreign antigen is detected by macrophage
   c. antigen recognition site on helper T-cells matches foreign antigen on macrophage
   d. antibodies are produced
   e. bacteria enter the body
   f. macrophage attacks invader

____ ____ ____ ____ ____ ____
First   Last

True or False Questions

Mark each question either T (True) or F (False).

____2. Tears have antibacterial qualities.

____3. Immunosuppressant drugs are likely to prevent infection by viruses.

____4. It is possible to carry the HIV virus without showing any symptoms.

____5. Stress may result in sickness due to lowered numbers of white blood cells.

____6. An antibody is very general in its effect, and is able to attach to many different types of bacteria.

Matching Questions

Each question may have more than one answer; write all of the answers that apply. Answers may be used more than once.

____7. inflammation

____8. macrophages

____9. mucous membranes

a. nonspecific defense response
b. specific response
c. both a and b
d. none of the above

____10. neutrophils

____11. antibodies

Multiple-Choice Questions

Circle the choice that is the best answer for each question.

12. Suppose that you cut yourself, and the next day there is swelling around the wound. The swelling is due to
    a. histamines.
    b. a secondary response.
    c. autoimmune reactions.
    d. antigens attacking the helper T-cells.

13. Helper T-cells are
    a. leukocytes.
    b. lymphocytes.
    c. white blood cells.
    d. all of the above.
    e. none of the above.

14. Cytotoxic memory cells are produced when _____ divide.
    a. macrophages
    b. helper T-cells
    c. B-cells
    d. interleukin molecules

15. During an autoimmune response the body's immune system begins to attack
    a. viruses.
    b. bacteria.
    c. interferon.
    d. body cells.

16. HIV can be transmitted by _____ an infected person.
    a. sitting in a classroom with
    b. shaking hands with
    c. visiting the hospital room of
    d. having unprotected intercourse with
    e. all of the above

Fill in the Blanks

Complete each statement by writing the correct word or words in every blank.

17. Antibodies are produced by __________ cells which are produced by activated __________ cells,

    that had been stimulated to divide by the chemical __________.

18 The pH of the body helps prevent infection due to the __________ environment of the stomach and the __________ environment of the intestine.

19. Interferon is produced by __________, to protect healthy cells from invasion by __________.

20. The HIV virus attacks __________ __________ cells and macrophages. This __________ the immune system.

Questions For Further Thought

Write your answer for each question in the space provided.

21. Explain why a vaccination may give you some (usually mild) symptoms of the disease, yet still provide long term protection.

SUMMARY

Study the fourteen items listed in the SUMMARY on p. 570 of the text. It should be apparent to you that there are several mechanisms in your immune system that are constantly at work to find and destroy viruses, cells that are not "self," and any other foreign bodies. These efforts are generally so successful that you are never aware of most of them, and you stay healthy.

You have also seen that the immune system relies on preventing the entry of invaders, as well as on generalized and specific short-term and long-term defenses once infection has occurred. The HIV virus presents unique problems because it destroys the cells which would normally be involved in controlling a viral infection.

HINTS

Read these suggestions before you check your answers to the EXERCISES.

For question3 : Remember, these drugs decrease the effectiveness of the immune system.

For question 6: This is asking about *one kind* of antibody.

For question 12: This is just one day later.

For question 13: Do not write "B" for your answer.

For question 16: HIV transmission requires an exchange of body fluids.

For question 17: The second answer is not "T-cells."

For question 21: Remember to consider both specific and nonspecific responses.

CONGRATULATIONS . . . YOU HAVE COMPLETED CHAPTER 20 !!!

CHAPTER 21

HORMONES AND NERVES

CONNECTIONS

In order to maintain homeostasis, the various regions of the body must communicate with one another. By communicating, they are able to monitor the conditions around the body and then stimulate any appropriate changes. This chapter examines the roles of hormones and nerves in coordinating these continuous processes. Look for ways that the body is able to exert specific controls.

Previous chapters have shown that living things are able to respond quickly to their environment through chemical activity and the actions of enzymes. Watch for these effects throughout this chapter as well. Additional aspects of the nervous system will be covered in the final two chapters of Part Five.

OBJECTIVES FOR THIS CHAPTER

When you have finished Chapter 21, you should be able to:

1. Describe some regulatory functions of invertebrate hormones.
2. Name and locate the major human endocrine glands.
3. List the hormones produced by each of the human endocrine glands and describe their effects.
4. Compare and contrast the mechanisms of action for both second messenger and steroid hormones.
5. Describe the three ways in which hormones and nerves are related.
6. Describe the structure and relationships of neurons, nerves, and glial cells.
7. Describe the three types of neurons, their impulse pathways, and physical characteristics.
8. Explain how an impulse is generated in and conducted along a neuron.
9. Discuss the role of neurotransmitters in synaptic transmission.

CHAPTER 21 OUTLINE

I. Hormones
  A. Invertebrate Hormones
II. Human Hormones

A. How Hormones Work
   1. Second Messenger Systems (Peptide Hormones)
   2. Direct Gene-Activation Systems (Steroid Hormones)
B. Feedback in Hormonal Systems

III. The Relationship of Hormones and Nerves

IV. Nerves
   A. Impulse Pathways
   B. The Mechanism of the Impulse
   C. The Synapse

CONCEPTS IN BRIEF

Hormones

Each different kind of hormone affects only certain cells, and is very specific in its action. Hormones are released (in very small quantities) from the cells of endocrine glands, and travel through blood vessels. When they reach their target cells (in another part of the body) they attach to specific receptor sites and begin a predictable series of changes in the cell. Depending on the hormone, these changes may be temporary or permanent.

Many types of organisms use regulatory hormones, but humans are emphasized in this chapter. Plants, insects, and crustaceans are also mentioned due to their extensive use of these chemicals for regulating growth, reproduction, homeostasis, and even for controlling their pests.

Human Hormones

With close to 200 hormones in the human body, the text focuses on only a few of these. Table 21.1 lists these, their points of origin, and the principle action in which each is involved. Although the pituitary, thyroid, and adrenal glands are responsible for secreting many of those listed, none of these glands can be said to be in charge of the body, since they all affect one another in the constant balance/counterbalance adjustments that are typical of homeostasis. As described in Chapter Eighteen, negative feedback controls are at work here.

Hormones affect what goes on inside cells. Second messenger systems accomplish this even though the hormone (called the first messenger) remains outside the cell. It attaches to a receptor site and triggers a series of reactions on the inside. The result is the formation of another chemical (the second messenger), commonly cyclic AMP, which causes a specific response by the cell.

Steroid hormones actually enter the cell and attach to the chromosome, along with a receptor molecule. Here it directs the production of specific mRNA molecules. As you recall from Chapter Six, the mRNA will then produce a protein, which in this case will cause a specific response by the cell.

The Relationship of Hormones and Nerves

Hormones and nerves interact in coordinating many of the body's activities. They are also related by the fact that some endocrine glands are formed from nervous tissue, and epinephrine (adrenaline) is found in both systems.

Nerves

Nerves initiate and conduct neural impulses along their nerve cells (neurons). These cells generally receive stimuli and begin their impulses in their dendrites. The impulses then go past the cell body, and pass through the axon on their way to another cell. Receptor cells receive stimuli from the environment, causing a sensory neuron to send an impulse to the spinal cord, where the message will be sent to the brain. Motor neurons carry messages from the brain back to the body. More details about these pathways can be found in Chapter Twenty Two.

Neural impulses differ from electrical impulses in several ways. The strength of the impulse remains the same as it travels the entire length of the neuron. Also, once it has fired, the neuron requires a very brief time (to repolarize) before it can fire again. Finally, the impulse in a neuron is always the same strength, regardless of the strength of the stimulus.

When a neuron is at rest it has an overall positive (+) charge on the outside of the membrane and a net negative (-) charge on the inside. If it is stimulated to fire, sodium ions (Na^+) rush inside and that part of the membrane shows a (-) charge on the outside and a (+) charge on the inside. This change is said to depolarize the membrane, causing an action potential to appear. Potassium ions (K^+) then rush out of the neuron, but the original ion distribution is quickly reestablished by the sodium-potassium pump, and the membrane is repolarized. All of this occurs very quickly, but it causes a domino affect of depolarizing/repolarizing to pass all along the length of the neuron. This "wave" of activity is what is called a neural impulse.

When the impulse reaches the end of the axon, there is no more cell membrane and it stops. In order to stimulate the next neuron to continue the impulse, a neurotransmitter is passed across the narrow gap (synapse), and stimulates the dendrites of the next neuron to fire. This substance is (very quickly) either destroyed by enzymes or reabsorbed by the original neuron, and the synapse is cleared to relay the next impulse.

KEY TERMS

Be sure to make and use flash cards for all of these terms. Suggestions are found in "To The Student" at the beginning of the Study Guide. Page numbers refer to the text.

acromegaly	576	first messenger	581
action potential	589	glial cell	587
adrenal gland	578	hormone	575
adrenaline	579	hypothalamus	576
afferent (sensory) neuron	587	impulse	587
aldosterone	578	inhibitory neurons	591
atrial natriuretic factor (ANF)	579	interneurons	587
axon	584	ion channels	589
calcitonin	576	myelin sheath	585
cell body	584	nerve	587
cortisol	578	neuron	584
cretinism	576	neurotransmitter	591
dendrite	584	norepinephrine	579
depolarization	589	parathyroid gland	576
effector	587	parathyroid hormone	576
efferent (motor) neuron	587	pituitary	576
endocrine gland	575	progesterone	578
epinephrine	579	receptor	587
estrogen	578	repolarization	589
excitatory neurons	591	resting potential	588
		saltatory propagation	585

second messenger	581	testosterone	578
sodium-potassium pump	589	threshold	589
somatotropin	576	thyroid	576
synapse	591	thyroxine	576
target organ	575		

EXERCISES

Check your understanding of Chapter 21 by completing the following exercises. Answers begin on p. 207.

Use the following diagram to answer Questions #1–7.

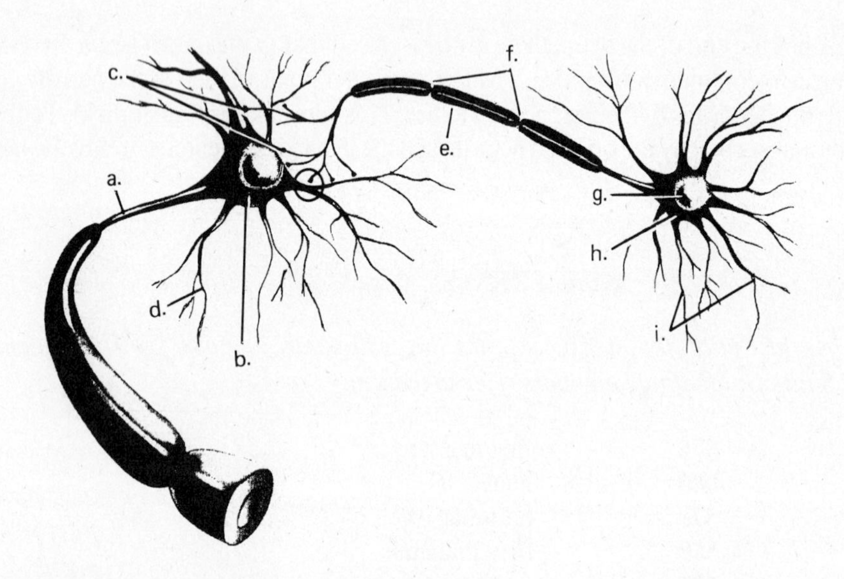

1. A neural impulse would normally travel from _____ in this diagram.
   a. right to left
   b. left to right
   c. either a or b

2. In saltatory propagation, an impulse leaps between the points indicated by letter(s)__________.

3. The presence of a cell body is indicated by letter(s) __________.

4. Letter(s) __________ give(s) the spinal cord its whitish appearance.

5. A neurotransmitter substance is found at letter(s) __________.

6. If these are responding to stimuli from the environment, the stimulation would occur at letter(s) __________.

7. An impulse in a neuron begins at letter(s) __________.

True or False Questions

Mark each question either T (True) or F (False).

____8. Hormones regulate some of the activities of cells in plants, arthropods, and mammals.

____9. Each endocrine gland operates on its own, regardless of what the other endocrine glands are doing.

____10. First messenger hormones enter the target cell at special receptor sites.

____11. Some endocrine glands are formed from nervous tissue.

____12. The strength of an action potential in a neuron depends on the strength of the stimulus.

Matching Questions

Each question may have more than one answer; write all of the answers that apply. Answers may be used more than once.

____13. excess sodium ions (Na^+) outside the neuron and excess potassium ions (K^+) inside

____14. K^+ ions rush out of the neuron

____15. Na^+ ions rush inside the neuron

a. action potential
b. resting potential
c. repolarization

____16. located in the brain

____17. hormones stimulate secondary sex characteristics

____18. regulates blood sugar levele. pancreas

____19. secretes growth hormone

a. anterior pituitary
b. adrenal glands
c. thyroid
d. gonads
f. none of the above

Multiple-Choice Questions

Circle the choice that is the best answer for each question.

20. Hormones are transmitted to other parts of the body through the
    a. axons.
    b. blood.
    c. homeostatic system.
    d. action potentials.

21. Steroid hormones regulate
    a. ATP usage during saltatory propagation.
    b. the rate at which neurons fire.
    c. the production of mRNA.
    d. how quickly second messengers are absorbed.

22. The neurons in a single nerve transmit their impulses
    a. in unison.
    b. independently of one another.
    c. independently of one another during weak stimulation, but in unison during strong stimulation.

23. Messages are sent from the brain to muscles along _____
    a. receptors.
    b. interneurons.
    c. afferent neurons.
    d. efferent neurons.

Fill in the Blanks

Complete each statement by writing the correct word or words in every blank.

24. The drawing which appears above Question #1 shows one complete __________ and part of another one.

25. Hormones are secreted by __________ glands.

26. A neural impulse is initiated in the __________ of a neuron.

27. A synapse is the place where a __________ substance passes from the __________ of one neuron to the __________ of another neuron.

28. The sodium pump in a neuron operates on energy from __________.

Questions For Further Thought

Write your answer for each question in the space provided.

29. Since an impulse can travel along a neuron in either direction, why do nerves only send impulses from the dendrite end to the axon end?

SUMMARY

Study the thirteen items listed in the SUMMARY on p. 594 of the text. You have seen that the chemical activities of hormones and nerves act to regulate and operate many of your body's activities. These two systems work together, and by using negative feedback, they provide ways for you to monitor and control what takes place throughout your body on a constant basis. In addition to the structural and mechanical differences in these two systems, you ought to be aware that communication via the nervous system is much more rapid than that which occurs due to the actions of hormones.

These actions take place in cells, so once again you are reminded that the activities of cells (and their chemistry) is essential to what your body as a whole does. You also should have noticed that most of what is discussed in this chapter are involuntary actions, which most people are not able to consciously control.

HINTS

Read these suggestions before you check your answers to the EXERCISES.

For question 1: The answer is *not* "c."

For question 9: Remember feedback control and coordination.

For question 10: The key word here is "enter."

For question 12: This is about a single neuron. Don't forget "all or nothing."

For question 21: Steroid hormones are part of the direct gene-activation system.

For question 22: A nerve is made up of many individual neurons. This is asking about the neurons themselves.

For question 28: Remember the energy molecule for almost all metabolic activities?

For question 29: Think about the role of neurotransmitter substances.

CONGRATULATIONS . . . YOU HAVE COMPLETED CHAPTER 21 !!!

CHAPTER 22

THE NERVOUS SYSTEM

CONNECTIONS

In the previous chapter, you examined the role played by the nervous system in communication throughout the body. Now you are about to be presented with a more detailed view of the structure of nervous systems, with an emphasis on that of a human and its many specialized components. Watch for the fact that different neural pathways are reserved for particular types of messages to or from specific parts of the body.

The brain and its functions are described here, as are ways in which people alter the normal activities of the brain through psychoactive agents. Look for the specialized functions (and interactions) of the regions on the brain. You will see which parts of the body are most sensitive to stimuli and where (in the brain) those sensations register. The next chapter will explain the various sense receptors of the body and their role in helping us interpret the world around us.

OBJECTIVES FOR THIS CHAPTER

When you have finished Chapter 22, you should be able to:

1. Describe the organization and complexity of representative invertebrate nervous systems.
2. Discuss the possible advantages of having a "brain" in the front end of an animal.
3. Describe the components and neural pathways of the central nervous system.
4. Explain how a reflex arc operates.
5. Discuss some trends in brain development among vertebrates, and relate these to the lifestyles of various animals.
6. List the major parts of the vertebrate brain and their functions.
7. Explain how the cerebral hemispheres obtain and process information.
8. ist the divisions of the peripheral nervous system and describe how each acts.
9. Give examples of psychoactive agents, their effects, and potential for dependence.

CHAPTER 22 OUTLINE

I. The Evolution of Nervous Systems
II. The Central Nervous System
    A. The Spinal Cord and the Reflex Arc
    B. The Vertebrate Brain
III. The Human Brain
    A. The Hindbrain
        1. The Medulla
        2. The Cerebellum and Pons
    B. The Midbrain
    C. The Forebrain
        1. The Thalamus and Reticular System
        2. The Hypothalamus
        3. The Cerebrum
    D. Hemispheres and Lobes
    E. Two Brains, Two Minds?
    F. The Split Brain
IV. The Peripheral Nervous System
    A. The Autonomic Nervous System
    B. Autonomic Learning
V. Mindbenders
    A. Tobacco
    B. Caffeine
    C. Marijuana and Hashish (THC)
    D. Alcohol
    E. Opiates
    F. Cocaine
    G. Amphetamines
    H. Barbiturates
    I. Phencylidine
    J. Methaqualone
    K. Psychedelics

CONCEPTS IN BRIEF

The Evolution of Nervous Systems

You may recall from Chapter Thirteen that the cnidarians were the first animals to show the development of true nerves. *Hydra* are shown here as an example of an animal with a nerve net, in which there are nerves throughout the body but there is no region of the nervous system which coordinates its actions.

Flatworms have a greater amount of coordination, and show the beginnings of a "brain" in the head. This location for the control center may allow an animal (that is moving forward) to sense and react to its environment more quickly.

Earthworms show the characteristics of a single (ventral) nerve cord running from one end of the animal to the other, containing numerous paired neurons which branch from the main cord and extend into the various regions of the body. This pattern is also evident in the vertebrates, however their spinal cord is on the dorsal side of the body and the brain is more completely developed.

The Central Nervous System

The central nervous system is composed of the brain and the spinal cord. Of course the rest of the nervous system communicates with these areas as well. Some responses, such as the knee-jerk response, occur automatically, without your conscious control. The stimulus triggers a response to an effector right in the spinal cord, and a reaction occurs. In addition, a message is sent to the brain and an additional, conscious response may follow. There are specific neurons along which each of these impulses travels.

Vertebrate brains have a medulla, cerebrum, and cerebellum, however these regions do not have the same relative importance in the different classes of vertebrates. In moving from fish, to reptiles, to birds, and to mammals, there are several trends which can be seen:

- an increase in the size of the brain relative to the rest of the body
- an increase in the size of the cerebrum relative to the rest of the brain
- generally higher intelligence.

There are also several important exceptions to each of these trends.

Mammals have a very highly developed cerebrum, which implies a reliance on intelligence for their survival. The mammals also exhibit an ability to learn (accompanied by parental care) and many examples of highly developed social life-styles.

The Human Brain

Your brain can be considered as consisting of three main regions: the hindbrain, midbrain, and forebrain. Table 22.2 describes these, their subdivisions, and their major functions. In general, the hindbrain controls many life sustaining activities along with balance and coordination. The midbrain is associated with input from the eyes and ears. Finally, the forebrain contains the regions of conscious thought and controls (such as speech, movement, intelligence, etc.), much sensory input, and communication and stimulation within the brain itself.

There are right and left halves (hemispheres) of your cerebrum, each of which can be divided into four lobes: the prefrontal, frontal, temporal, and occipital. Electrical probing has revealed that specific brain functions are found in predictable regions of these lobes. It has also been shown that the most sensitive regions (in terms of sensory receptors) of the body are the face, hands, and genitals. You can exert the greatest amount of control over your face and hands. Notice that these two lists are not exactly the same.

Current evidence indicates that the left hemisphere of the cerebrum may exert more control over logical thought and language. The right hemisphere may specialize in more artistic and imaginative activities. The two sides are connected by (and communicate via) the corpus callosum. As long as this connection operates, some of these functions overlap. Visual input from each eye goes to both hemispheres via the optic chiasma.

The Peripheral Nervous System

Go back to Table 22.1 and look at the components of the peripheral nervous system. These are the nerves which branch into the rest of the body from the central nervous system. Notice that the peripheral nervous system is subdivided into the somatic nervous system (concerned with voluntary movement) and the autonomic nervous system, which deals with involuntary activities. This section of the text emphasizes the actions of the autonomic nervous system.

The autonomic nervous system is further subdivided into the sympathetic nervous system and the parasympathetic nervous system. Your sympathetic nervous system is involved in reproduction and in preparing your body to respond (immediately and emphatically) to emergency situations. The parasympathetic nervous system is responsible for controlling numerous organs under normal circumstances, as well as for the reactions which return your body systems to normal after an emergency is over. The neurons in these two systems may go to the same organs, but they utilize different neurotransmitter substances.

It is clear that some people can control body functions that had been thought to be entirely in the realm of involuntary control. The regulation of metabolism, and the ability to withstand severe heat and cold have been seen, but our research on these has not revealed the mechanism behind these unexpected results.

Mindbenders

Throughout history there have been numerous ways in which people have altered their brain functions by using legal and illegal psychoactive agents. The main effects of these are either: (1) some degree of stimulation or depression of the brain, or (2) a hallucinogenic effect from psychedelics. Table 22.3 summarizes many of the chemicals which are in current use in the world, including their long and short term actions and effects. Many of these drugs produce some level of physiological or psychological dependence in the users, which may result in withdrawal reactions if they stop taking the drug. People may also develop a tolerance to the effects of certain of these drugs, thereby requiring higher dosages to achieve the same effects.

KEY TERMS

Be sure to make and use flash cards for all of these terms. Suggestions are found in "To The Student" at the beginning of the Study Guide. Page numbers refer to the text.

Term	Page
acetylcholine	616
addiction	622
ascending neuron	601
biofeedback	619
caffeinism	625
central nervous system	600
cerebellum	602 & 607
cerebrum	602 & 609
corpus callosum	614
cranial nerve	615
depressant	621
descending neuron	601
dorsal root	600
forebrain	605
frontal lobe	609
hashish	625
hindbrain	605
hypothalamus	609
interneuron	601
marijuana	625
medulla	602 & 605
midbrain	605
narcotic	628
norepinephrine	616
occipital lobe	609
optic chiasma	614
parasympathetic nervous system	616
parietal lobe	610
peripheral nervous system	614
physical dependence	622
pons	605 & 607
prefrontal area	610
psychedelic	632
psychoactive agent	621
psychological dependence	622
reflex arc	600
reticular system	607
sedative-hypnotic	631
spinal nerve	615

EXERCISES

Check your understanding of Chapter 22 by completing the following exercises. Answers begin on p. 207.

Use the following drawings to answer Questions #1–6.

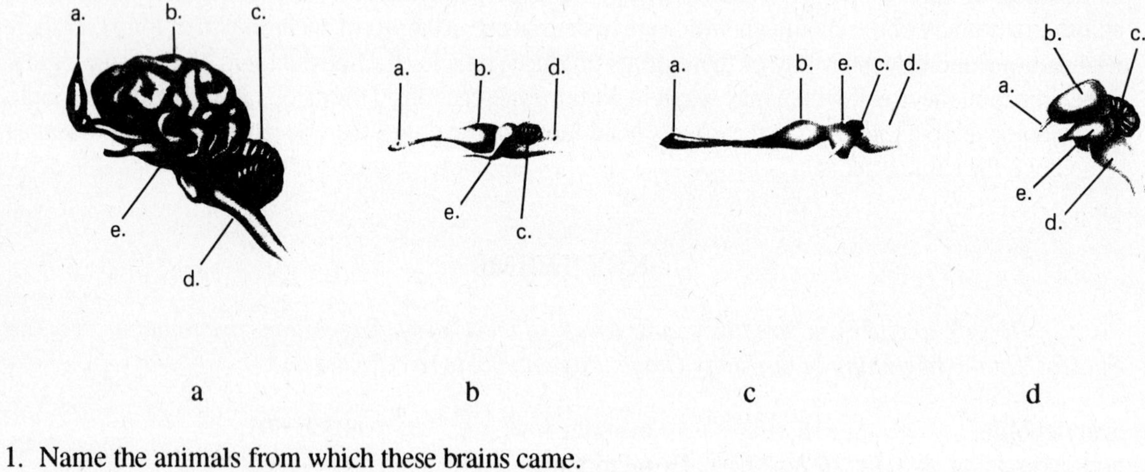

1. Name the animals from which these brains came.

   a. ____________________ b. ____________________

   c. ____________________ d. ____________________

2. Structure B is the __________.

3. Animal _____ would rely the least on its sense of smell.

4. Structure _____ controls balance.

5. Animal _____ seems to have the greatest emphasis on hindbrain control.

6. Animal _____ probably has the most emphasis on forebrain control.

True or False Questions

Mark each question either T (True) or F (False).

____7. All vertebrates have a central nervous system.

____8. A reflex arc is controlled by the cerebrum.

____9. An increase in size of the cerebrum means an increase in intelligence.

____10. The two hemispheres of the brain are identical in their functions.

____11. The parasympathetic nervous system is used as a backup system for the sympathetic nervous system in emergencies in order to provide higher levels of response.

____12. Tobacco smoke is harmful to your lungs, but marijuana smoke is not harmful to them.

____13. The effects of barbiturates can be counteracted by drinking alcohol.

Matching Questions

Each question may have more than one answer; write all of the answers that apply. Answers may be used more than once.

____14. neurons which cause muscles to move

____15. pons is part of this

____16. brings about recovery after an emergency reaction

____17. contains ascending neuron

____18. location of memory

a. central nervous system
b. peripheral nervous system
c. autonomic nervous system
d. somatic nervous system
e. parasympathetic nervous system
f. sympathetic nervous systems
g. all of the above
h. none of the above

Multiple-Choice Questions

Circle the choice that is the best answer for each question.

19. The _____ is (are) the region(s) of the body most sensitive to stimuli.
    a. genitals
    b. mouth
    c. hands
    d. all of the above
    e. none of the above

20. Impulses sent from your right eye are sent to
    a. the right side of your brain.
    b. the left side of your brain.
    c. both sides of your brain.

21. The responses which are triggered by the autonomic nervous system are
    a. completely involuntary.
    b. completely voluntary.
    c. mostly involuntary, but some may be controlled consciously.
    d. mostly voluntary, but some may be controlled consciously.

22. Cocaine affects your body as a(n)
    a. addictive stimulant.
    b. addictive depressant.
    c. nonaddictive stimulant.
    d. nonaddictive depressant.
    e. hallucinogen.

Fill in the Blanks

Complete each statement by writing the correct word or words in every blank.

23. The three main regions of your brain are the __________ in front, the __________ in the middle, and the __________ at the rear.

24. The __________ system regulates whether or not certain impulses will register in your brain.

25. Caffeine is a (an) __________ to the central nervous system, but alcohol is a (an) __________.

26. The autonomic nervous system is made up of the __________ nervous system and the __________ nervous system.

Questions For Further Thought

Write your answer for each question in the space provided.

27. What are some advantages to having reflex arcs?

SUMMARY

Study the ten items listed in the SUMMARY on p. 634 of the text. It should be clear to you that the human nervous system can be divided into two major divisions (with several subdivisions), and that each of these is responsible for specific actions. Some responses function with no conscious control, some respond to your "decision making," and some can be influenced by both. Overall, the nervous system delivers very rapid responses to stimuli from the environment, which provides the organism with an additional characteristic for increased chances of survival.

Although a great deal is known about the brain, there is much that is not known about communication within the brain, memory, sensitivity to stimuli, and other aspects of the central nervous system. The reactions within and between neurons are electrochemical in nature, so it is not surprising that numerous chemicals can alter the responses within the nervous system.

HINTS

Read these suggestions before you check your answers to the EXERCISES.

For question 1: Look at Figure 22.3.

For question 8: An arc does not go to the brain.

For question 9: This considers overall size, all of the time.

For question 10: Don't forget about right and left brain effects.

For question 13: They are both depressants.

For question 16: Consider the subdivisions of the nervous system. There is more than one answer to this question.

For question 25: See Table 22.3 for the mode of action.

CONGRATULATIONS . . . YOU HAVE COMPLETED CHAPTER 22 !!!

CHAPTER 23

THE SENSES

CONNECTIONS

An organism's perception of the world is dependent on what its sense receptors detect and how it interprets that information. The last chapter described how environmental stimuli travel to the brain and register there. Chapter Twenty-Three examines six types of receptors and their methods of responding to the environment. Look for the specialized nature of these receptors and their relationships to the nervous system.

Part One of the text described the effects of variation and natural selection on populations of organisms. This chapter provides several examples of how an individual organism's senses are used for its own survival. Think about how different kinds of receptors may interact as you study this material. Be sure to apply this information to both individuals and populations. The next section of the text will explore these ideas further.

OBJECTIVES FOR THIS CHAPTER

When you have finished Chapter 23, you should be able to:

1. Describe six main types of sensory receptors and the stimuli to which they respond.
2. Compare and contrast the function and distribution of vertebrate and invertebrate receptors.
3. Discuss the role of each of the six types of receptors in the survival of invertebrates and vertebrates.
4. Name the structures of the vertebrate ear and explain how each functions to convert sound waves to neural impulses.
5. Describe the various types of chemoreceptors.
6. Identify four different tastes and indicate where each is registered.
7. Discuss the mechanisms involved in maintaining balance.
8. Identify the main structures in the eye and describe their functions.
9. Compare and contrast the functions of rods and cones in the vertebrate eye.

CHAPTER 23 OUTLINE

I. Thermoreceptors
II. Tactile Receptors
III. Auditory Receptors
IV. Chemoreceptors
V. Proprioceptors
VI. Visual Receptors

CONCEPTS IN BRIEF

Thermoreceptors

The names of each of the types of receptors tells you what they detect (remember *thermo*dynamics?). A mobile animal which is sensitive to heat can alter its behavior to escape dangerous conditions and put itself in a more optimal environment. Heat sensitivity may also be used to find food. Humans are sensitive to changes in temperature, and register extremes of hot or cold as pain.

Tactile Receptors

Touch receptors are sensitive to movement, especially of muscles and hairs. Animals commonly have certain regions of their body which are more sensitive than others. Primates have concentrations of receptors in their genitals, lips, eyes, fingers, and hairy areas. This should remind you of the material in Chapter Twenty-Two.

Auditory Receptors

Sound travels as vibrations in air, water, and soil. Insects are the best of the invertebrates at detecting these vibrations. Mammals generally have an external ear, in addition to the auditory structures found in other vertebrates.

Air movement causes the human eardrum to vibrate and move (in the following order): three bones, the oval window of the inner ear, fluid inside the cochlea, the basilar membrane, and hair cells. When the hair cells move (which ones and how many are important), impulses are sent to the brain and interpreted as sound.

Chemoreceptors

Taste and smell are discussed here, as responses to specific chemicals. Animals are able to detect some of these from long distances, which gives them more of an opportunity to respond properly. Many insects are very sensitive to minute amounts of substances in the environment.

Vertebrates require that materials be dissolved and move across moist membranes before they can be detected. Human smell receptors (in the nasal passage) are directly connected to the olfactory lobe of the brain. Taste receptors are found in the tongue, and are sensitive to four basic tastes: sweet, sour, salty, and bitter.

Proprioceptors

You maintain your balance and monitor your body position with input from your proprioceptors. Balance receptors, which are sensitive to the effects of gravity, are found in the semicircular canals of the ears.

Visual Receptors

Animals have visual pigments which produce neural impulses when exposed to light. Humans use visual purple in the retina for this. You see color because of the cones in the retina, but they require more light than the rods, which are not sensitive to color.

KEY TERMS

Be sure to make and use flash cards for all of these terms. Suggestions are found in "To The Student" at the beginning of the Study Guide. Page numbers refer to the text.

audition	640	proprioceptor	650
auditory canal	640	receptors	639
basilar membrane	645	retina	651
chemoreception	646	rods	651
cochlea	645	saccule	650
cones	651	semicircular canals	650
external ear	641	stapes	645
gustation	646	tactile receptors	640
incus	645	tectorial membrane	645
malleus	645	thermoreception	639
Meissner's corpuscles	640	tympanal membrane	640
middle ear	645	tympanic membrane	640
olfaction	646	utricle	650
olfactory bulb	647	visual pigment	651
oval window	645	visual purple	651
Pacinian corpuscles	640		

EXERCISES

Check your understanding of Chapter 23 by completing the following exercises. Answers begin on p. 207.

1. Arrange the following events of auditory reception in sequence from first to last.
   a. stapes vibrates
   b. basilar membrane moves
   c. air vibrations pass the pinna
   d. oval window moves
   e. cochlear fluid moves
   f. eardrum vibrates

____ ____ ____ ____ ____ ____
First   Last

True or False Questions

Mark each question either T (True) or F (False).

____2. The semicircular canals are important for detecting differences in the pitch of sound.

____3. Only mammals are able to see true colors.

____4. Some insects are able to detect sound.

____5. Humans use different receptors for light touches and pressure.

Matching Questions

Each question may have more than one answer; write all of the answers that apply. Answers may be used more than once.

Use the letters to indicate what each of the following is sensitive to in the environment.

____6. auditory receptors

____7. tactile receptors

____8. thermoreceptors

____9. visual receptors

a. vibrations
b. portions of the electromagnetic spectrum
c. changes in heat
d. hair movement
e. none of the above

Multiple-Choice Questions

Circle the choice that is the best answer for each question.

10. Whiskers of vertebrates are generally supplied with
    a. auditory receptors.
    b. chemoreceptors.
    c. tactile receptors.
    d. proprioceptors.
    e. visual receptors.

11. Humans are able to detect many different kinds of tastes because
    a. they have many different kinds of taste buds in the tongue.
    b. the colors of the foods stimulate many different kinds of taste neurons to fire.
    c. different foods result in different patterns of impulses from the taste buds.
    d. different foods stimulate the four kinds of taste buds along with many different kinds of endocrine gland secretions.

12. The light-sensitive region of the eye is the
    a. lens.
    b. cornea.
    c. pupil.
    d. retina.
    e. vitreous humor.

13. The _____ are influenced by gravity.
    a. auditory receptors
    b. chemoreceptors

c. proprioceptors
d. thermal receptors

Fill in the Blanks

Complete each statement by writing the correct word or words in every blank.

14. There are several types of neural structures called __________, which are able to respond to stimuli from the environment.

15. The tip of your tongue is most sensitive to __________, and the rear part is sensitive to __________.

16. Human thermoreceptors provide sensations of __________, __________, and __________.

17. The receptors which would be stimulated by stretching and bending are your __________.

Questions For Further Thought

Write your answer for each question in the space provided.

18. Explain why colors disappear from your vision as light becomes more and more dim.

19. Discuss possible reasons why noctuid moths are sensitive to sound volume but not sensitive to different frequencies of sound.

SUMMARY

Study the eight items listed in the SUMMARY on p. 653 of the text. You have seen that organisms show variation in their sensitivity to environmental stimuli. This variation relates to the life-style of each type of animal, as you would expect.

HINTS

Read these suggestions before you check your answers to the EXERCISES.

For question 2: These are part of the proprioceptor system.

For question 6: This has more than one answer.

For question 12: The key word here is "sensitive."

For question 14: This is a general term.

For question 18: What part of your eye is sensitive to color?

CONGRATULATIONS . . . YOU HAVE COMPLETED CHAPTER 23 !!!

Part Six Behavior and the Environment

CHAPTER 24

ANIMAL BEHAVIOR

CONNECTIONS

It is very common for people to talk about other animals in human terms, and assume that their actions are an attempt to act like humans. This chapter takes a more scientific approach in examining the components of certain animal behaviors and the reasons for their existence. Like other characteristics, behavior changes through time and is subject to the effects of natural selection (including reproductive success). Look for references to these as you read.

There is some disagreement over how much of an animal's behavior is the result of (or is influenced by) its genes, so you will have another chance to apply the concepts of genetics here. In addition, behaviors are certainly influenced by the environment and the stimuli which are presented to an animal's senses. As you can see, many of the ideas which have been introduced in earlier chapters are tied together now, as you begin Part Six of the text. In subsequent chapters, these animal behaviors will be applied in the even wider context of the planet and our relationship to it.

OBJECTIVES FOR THIS CHAPTER

When you have finished Chapter 24, you should be able to:

1. Compare and contrast the ways in which ethologists and comparative psychologists study animal behavior.
2. Describe the components of instinctive behavior.
3. Give examples of three major types of learning and explain the components of each.

4. Explain the relationship between learning and innate behavior.
5. Describe imprinting and its effects.
6. Explain how birds orient and navigate.
7. Describe how animals interact through aggression.
8. Describe how animals interact through cooperation.
9. Give examples of altruistic behaviors and explain why they are adaptive.
10. Describe how sociobiologists view human behavior and explain why sociobiology is controversial.

CHAPTER 24 OUTLINE

I. Ethology and Comparative Psychology
II. The Development of the Instinct Idea
  A. Releasers
  B. Innate Releasing Mechanisms
  C. Fixed Action Patterns
III. Learning
  A. Habituation
  B. Classical Conditioning
  C. Operant Conditioning
IV. How Instinct and Learning Can Interact
  A. Imprinting
V. Orientation and Navigation
VI. Social Behavior
  A. Aggression
  B. Fighting
VII. Cooperation
VIII. Altruism
  A. Kin Selection
  B. Reciprocal Altruism
IX. Sociobiology and Society

CONCEPTS IN BRIEF

Ethology and Comparative Psychology

Throughout the chapter you will find references to two ways of approaching animal behavior. Ethology was developed in Europe, and emphasizes studying animals in their natural environments. Comparative psychology began in the United States and concentrated on experimenting with laboratory animals in controlled conditions. Many researchers use components of both approaches.

The Development of the Instinct Idea

An instinct is a behavior which has very specific actions (fixed action patterns). There are several additional components:

- it has a genetic basis
- it requires a particular stimulus (the releaser) from the environment
- it occurs after part of the central nervous system (the innate releasing mechanism) responds to the stimulus.

Notice that this is more restrictive than the way in which you may use the term "instinct" in general conversation.

Learning

Many animals learn a wide range of behaviors which are important to their own survival. It is not meaningful to compare what one species or another can be taught by humans. Higher intelligence, as we define it, is more common in social animals that have long life spans and which mature slowly. Three ways of learning are presented here: habituation, classical conditioning, and operant conditioning.

Habituation occurs when an animal learns to ignore a stimulus. Consequently the nervous system is not always overloaded by responding to *everything*. This learning is not permanent.

By exposing an animal to classical conditioning, it can learn to show a regular response when subjected to a stimulus which is different from the normal stimulus. If the ordinary reinforcement does not follow the new stimulus for several trials, the response will no longer be shown. Pavlov's experiments used this method.

Operant conditioning involves waiting until an animal shows a behavior which the researcher wants it to learn; then the animal is reinforced. Experiments done in a Skinner box are examples of this technique.

How Instinct and Learning Can Interact

It is clear that many behaviors are the result of innate patterns which have been modified or refined through learning and practice. Konrad Lorenz found that some species show imprinting. They have a specific period of time (early in life) in which they respond to their environment and are affected for the rest of their lives. If the normal environment is changed during the critical period, so is the resultant behavior.

Orientation and Navigation

Birds are able to use a variety of clues from the environment when they fly from one point to another, as well as when they choose to face one way or another. These include: the sun, the stars, physical features on the ground, magnetic fields, smell, sonar, and internal biological clocks.

Social Behavior

When two animals are competing for the same thing in the environment there may be conflict. Any interaction in which one animal tries to dominate the other is called aggression. This is most common among individuals of the same sex and species, since they are the most likely to be sharing the environment and be competing.

More fighting may occur among different species than has been realized in the past. There are organisms other than humans which may kill members of their own species, especially when placed under stressful conditions.

Ritualized combat between members of the same species results in fewer serious injuries, and the loser is generally allowed to escape. The conflict is still able to be solved, and the result may be less frequent fighting in the future.

Cooperation

Cooperation is another behavior which increases the chances that an individual will survive to reproduce. It is seen between members of different species (you may once have incorrectly defined it as mutualism) as well as members of the same species (especially social animals). It is commonly used for acquiring food and protection. Honey bees are used as an example of highly developed insect cooperation.

Altruism

Altruism occurs when an animal puts itself at risk for the benefit of another. On the surface this would not seem to present any advantage for the animal at risk, however think back to the idea of reproductive success (fitness) from Chapter One. If altruism increases the chance that individuals which share some of the same genes will survive, it may be advantageous to all the animals involved. The more closely related the animals are, the more likely it is that altruism will occur. It is assumed that this characteristic arose through natural selection, and not by any conscious choice of the animals involved.

Social, intelligent animals that can plan into the future may exhibit reciprocal altruism, in which there *is* a choice to help another. The choice assumes some future benefit to the individual that provides the help.

Sociobiology and Society

Edward O. Wilson has promoted the formalized study of the relationship of natural selection and social behavior (sociobiology). Critics have suggested that sociobiologists are saying that groups of people (races, sexes, etc.) are genetically programed to act in certain ways. This would imply that fundamental social change isn't possible, which could give scientific support to racism and sexism. Sociobiologists disagree, and support unbiased research as a way to expand our knowledge.

KEY TERMS

Be sure to make and use flash cards for all of these terms. Suggestions are found in "To The Student" at the beginning of the Study Guide. Page numbers refer to the text.

aggression	673	instinct	660
altruism	683	intelligence	663
biological determinism	688	interspecific	675
classical conditioning	665	intraspecific	680
comparative psychology	659	kin selection	684
cooperation	680	learning	662
critical period	669	navigation	
ethology	659	operant conditioning	666
extinction	666	orientation	670
fixed action pattern	661	reciprocal altruism	687
habituation	663	releaser	660
imprinting	669	Skinner box	666
innate behavior	660	sociobiology	688
innate releasing mechanism (IRM)	661		

EXERCISES

Check your understanding of Chapter 24 by completing the following exercises. Answers begin on p. 207.

1. Place the following aspects of innate behavior in sequence from first to last.
   a. innate releasing mechanism is stimulated
   b. releaser is perceived
   c. fixed action pattern occurs
   d. impulses are sent from central nervous system to muscles
   e. environmental stimulus exists

_____ _____ _____ _____ _____
First   Last

True or False Questions

Mark each question either T (True) or F (False).

____2. Circus elephants and horses which are able to perform tricks are more intelligent than species which are not used in the circus.

____3. An ethologist would spend most of his or her research time in a laboratory.

____4. It is not common to find biological rhythms in land animals.

____5. Sociobiologists generally say that a person's ancestry can be used to predict their behavior.

____6. If something is learned by habituation it is permanent.

Matching Questions

Each question may have more than one answer; write all of the answers that apply. Answers may be used more than once.

____7. altruism

____8. aggression

____9. cooperation

____10. kin selection

a. an interspecific behavior
b. an intraspecific behavior

____11. a Skinner box

____12. rewarding a behavior after the behavior occurs

____13. Pavlov's experiments with salivating dogs

a. classical conditioning
b. operant conditioning

Multiple-Choice Questions

Circle the choice that is the best answer for each question.

14. In classical conditioning, the _____ occurs before the _____.
    a. reinforcement, stimulus
    b. response, reinforcement
    c. stimulus, response
    d. fixed action pattern, stimulus
    e. none of the above

15. Practicing _____ behaviors may improve how well the behavior is performed.
    a. learned
    b. innate
    c. both a and b
    d. none of the above

16. The animals which show the most highly developed levels of cooperation are the
    a. insects.
    b. birds.
    c. primates.
    d. dogs and wolves.

17. A specific act of altruism in squirrels would be more likely to occur between
    a. a male and a female.
    b. a female and an unrelated neighbor.
    c. a parent and its offspring.
    d. all of the above are equally likely if the risk in each is the same.

18. If caged birds are deprived of sunlight and exposed to an artificial light (in a position which is different from the sun's position), they will orient toward
    a. the sun.
    b. the artificial light.
    c. the sun during the day and the artificial light at night.
    d. the sun on sunny days and the artificial light during cloudy days.

Fill in the Blanks

Complete each statement by writing the correct word or words in every blank.

19. Three ways in which animals learn are operant __________, __________ conditioning, and __________.

20. Aggression is usually associated with __________ for something in the environment.

21. Some students (not you, of course) study in noisy places. If they have learned to ignore the noise it would be an example of learning by __________.

22. In ritualized fighting the loser generally ________.

Questions For Further Thought

Write your answer for each question in the space provided.

23. a. Describe how it was possible for Konrad Lorenz to get ducklings to follow him as he walked around.

    b. Explain why they didn't learn to ignore him.

SUMMARY

Study the eleven items listed in the SUMMARY on p. 690 of the text. You should see that the study of animal behavior involves the use of a variety of research techniques. These are used to examine the relationships between stimuli and animal responses, and the many factors that affect these.

It should also be clear that many adaptive behaviors are the result of the interaction of genetics and learning, even though the mechanisms may not yet be understood. The genetic component of a species' behavior is affected by natural selection, as is any genetic characteristic. The learned component of an individual's behavior may be changed by the conditions of the environment. Whether or not a species learns a specific behavior in its natural environment is closely related to how useful that behavior is for survival and reproduction.

HINTS

Read these suggestions before you check your answers to the EXERCISES.

For question 2:	Intelligence is more than learning tricks.
For question 17:	This question could apply to any animal, not just squirrels.
For question 20:	It occurs when two organisms are trying to use the same resource.
For question 22:	They don't usually die.

CONGRATULATIONS . . . YOU HAVE COMPLETED CHAPTER 24 !!!

CHAPTER 25

BIOMES AND COMMUNITIES

CONNECTIONS

Throughout the text you have examined the structures and activities of organisms. You have considered the forces that affect these living things and the ways in which populations change through time. This chapter looks at a broader view. It is now time for a global perspective, and to consider the interactions of large groups of organisms with their environments.

The regions of the earth can be divided by their physical characteristics and by the life forms which inhabit them. You will see that there are numerous criteria for these divisions. Also watch for the difficulties in defining boundaries between divisions. Remember, nature often shows us a continuum of features, and in this chapter you will see that they *all* interact with one another.

OBJECTIVES FOR THIS CHAPTER

When you have finished Chapter 25, you should be able to:

1. List the components of an ecosystem.
2. Distinguish between a habitat and a niche.
3. Describe ways in which different species may coexist within a habitat.
4. Distinguish between primary and secondary succession.
5. Describe how communities change through primary and secondary succession.
6. Explain the role of producers in food chains and food webs.
7. List the types of consumers within a food chain.
8. Explain why the feeding patterns of organisms are best described as food webs.
9. Explain ways in which species extinction may affect the balance of life at different trophic levels.
10. Describe the characteristics of six major types of terrestrial biomes.
11. Compare and contrast freshwater and marine environments and their communities.
12. Describe the types of coastal environments and their communities.
13. Discuss the causes and possible effects of current rates of extinction on the earth today.

CHAPTER 25 OUTLINE

I. Ecosystems and Communities
II. Habitat and Niche
III. Succession
IV. The Web of Life
  A. Extinction and the Web of Life
  B. Extinction and Us
V. The Land Environment
  A. Biomes
VI. The Water Environment
  A. Freshwater Bodies
    1. Rivers and Streams
    2. Lakes and Ponds
  B. The Oceans
VII. Coastal Areas

CONCEPTS IN BRIEF

Ecosystems and Communities

You should recognize species and populations, as they were discussed in previous chapters. A group of populations in one place (you can define the boundaries yourself) which interact are a community. If you include the nonliving factors of the environment in which the community (or even several communities) lives, it is all called an ecosystem. Scientists study these different levels both independently and as interacting units.

Habitat and Niche

Both habitat and niche are associated with individuals. Habitat deals with location, and the features of that place. Niche is a broader term that includes habitat <u>and</u> all of the interactions of the individual, with both living and nonliving things. Robert McArthur's warbler study supports the idea that, even though many organisms may share a habitat, they do not share the same niche (at least not for long).

Succession

It is natural for communities to change over time. In fact they may change in predictable ways until they reach the climax stage, which is generally stable (as long as the environment remains stable as well).

If communities develop in an environment that does not have (and never has had) living things, we describe it as primary succession. Pioneer organisms are important in beginning this process. If, however, previous communities were disturbed (or destroyed), the process of reestablishing new communities is called secondary succession.

The Web of Life

Can you remember the laws of thermodynamics from Chapter Four? If so, you know that all living things require energy, and must periodically replenish their supplies of it. Food chains illustrate the relationships in this transfer of energy by showing who eats whom. Of course it starts with producers, and each level is named.

Food webs are a more realistic description of the complexity of interactions in real ecosystems. They show how food chains overlap and interlock. Just because they may not be as visible, don't forget the vital role played by the decomposers.

As a system becomes more complex it is also more likely to be able to withstand various types of disturbances. It is more likely to continue to function, and its members are less likely to become extinct.

Humans, however, can cause severe environmental change in very short periods of time, which has resulted in increased rates of extinction worldwide. Because of the fact that all living things are interconnected, the short- and long-term results of these extinctions are often difficult to predict. Ecologists feel that the loss of biodiversity poses dangerous possibilities for humans, and the rest of the earth.

The Land Environment

Large areas of land which support specific prominent types of plant growth are described as biomes. There are predictable types of animals in any biome, which influence the plant growth (which influence the animals, etc.). Environmental factors set the stage for which plant types dominate the area.

The text describes six major biomes: tropical rain forest, temperate deciduous forest, taiga, tundra, grassland, and desert. A summary of each of these and their characteristics can be found in Table 25.3 of the text. Notice the influence of climate in determining which biomes appear in the various parts of the world. You should be able to describe the characteristics of each biome, including its relative diversity.

The Water Environment

Both fresh water and salt water have salt in them, they just vary in how much each contains (salt water has more). Rivers and streams carry fresh water which is moving in one direction. As the water runs downstream it generally slows down and acquires more and more nutrients and silt. In addition, the species composition will change.

The water in lakes and streams is more still, and has a greater chance to accumulate materials, rather than carry them away (as is done by rivers). Fertilizers and detergents have caused rapid eutrophication of many lakes and ponds, which causes them to age more quickly.

Oceans are very large, but like all ecosystems they have limits to the amount of trauma they can withstand. The most biologically productive areas of the oceans are the shallow waters (generally near the shores) of the neritic province, where there is sufficient light for photosynthesis. Photosynthetic phytoplankton in these shallow waters produce a large portion of the earth's oxygen. Most food chains in the oceans begin with phytoplankton, although some center around chemosynthetic organisms in deep thermal vents.

Organisms which live at great depths must accommodate the tremendous water pressure by equalizing it inside and outside their bodies. Much research still needs to be done before we will have a better understanding of the deep communities.

Coastal Areas

Coastal waters are *the* most productive areas of the oceans. They contain more different kinds of living things than are found in open water, and are important reproductive habitats. They can be classified by their physical features into: rocky coasts, sandy seashores, and mud flats. Wave action and tidal movement are important in influencing what lives in the different types of coastal areas.

KEY TERMS

Be sure to make and use flash cards for all of these terms. Suggestions are found in "To The Student" at the beginning of the Study Guide. Page numbers refer to the text.

biomes	695	pioneer organisms	698
climax stage	698	populations	695
community	695	primary consumers	701
competition	697	primary succession	698
continental shelf		producers	
decomposers	703	quaternary consumers	701
deserts		rocky coasts	730
ecology	695	sandy seashores	731
ecosystem	695	secondary consumers	701
estuary		secondary succession	699
eutrophication	725	succession	698
extinction	704	taiga	712
food chain	726	temperate deciduous forest	712
food web	701	tertiary consumers	701
grasslands	716	trophic level	701
habitat	696	tropical rain forests	710
jungles	710	tundra	713
mud flats	731	zooplankton	726
niche	696		
phytoplankton	726		

EXERCISES

Check your understanding of Chapter 25 by completing the following exercises. Answers begin on p. 207.

1. Arrange the following terms in sequence from most simple to most complex.
   a. species
   b. ecosystem
   c. population
   d. individual
   e. community

   _____ _____ _____ _____ _____
   Simple Complex

Use Figure 25.30 in the text to answer Questions #2–4.

2. Because they are photosynthetic, the phytoplankton are also known as_________.

3. The next level up is composed of _________. They are the primary _________.

4. The shark represents a _________ level known as the _________ _________.

True or False Questions

Mark each question either T (True) or F (False).

_____5. Sandy shores have fewer kinds of living things than mud flats.

_____6. An organism's niche is one part of its habitat.

_____7. A complex community is said to be more stable than a simple community.

_____8. The rate of species extinction is now lower than it was one hundred years ago.

_____9. The taiga commonly has few plants growing on the forest floor.

_____10. Succession is a process by which communities remain the same over time.

Matching Questions

Each question may have more than one answer; write all of the answers that apply. Answers may be used more than once.

_____11. have the highest salt content

_____12. contain only fresh water

_____13. are the most likely to show eutrophication

a. estuaries
b. lakes
c. rivers
d. rocky shores

Multiple-Choice Questions

Circle the choice that is the best answer for each question.

14. The living things in tropical rain forests have _____ diversity.
    a. no
    b. very low levels of
    c. moderate levels of
    d. high levels of

15. Places which contain tundra are areas with _____ growing seasons.
    a. no
    b. short
    c. approximately six month long
    d. almost year long

16. The coastal areas with the least diversity of organisms are the
    a. mud flats.
    b. rocky shores.
    c. sandy seashores.
    d. jungle shorelines.

17. Nutrients are returned to the environment by the actions of the
    a. zooplankton.
    b. decomposers.
    c. herbivores.
    d. tertiary consumers.

18. The _____ biome is commonly found just above and below tropical rain forests on the globe.
    a. desert
    b. tundra
    c. taiga
    d. polar ice cap

Fill in the Blanks

Complete each statement by writing the correct word or words in every blank.

19. In the ocean you will find deeper water in the __________ province, and shallower water in the __________ province.

20. Florida is in the ____________________ biome, and most of southern Canada is in the ____________________ biome.

21. It is common for the ____________________ biome to occur along the equator.

22. If the vegetation of a riverbank is all washed away in a flood, new communities would become established by __________ succession.

23. Temperatures generally __________ as you move from the equator toward the poles, and __________ as you gain altitude.

Questions For Further Thought

Write your answer for each question in the space provided.

25. Look at Figure 25.14 in the text. According to this, why does chaparral grow in some places, but savanna grows in others?

a. What concept was supported by the data from McArthur's warbler study?

b. Explain why the data supported that concept.

SUMMARY

Study the fifteen items listed in the SUMMARY on p. 732 of the text. You should see that ecologists study environments on many different levels, but the common theme throughout their studies is the interaction of living things with each other, and with the environment.

Chapter One's accounts of Darwin (and his concept of how the environment affects organisms) should have come to mind as you kept reading (in this chapter) of the strong influence of the environment on the types of living things which live in any habitat. Introduced here was the additional idea that living things also exert an influence on the environment by their presence and activities.

Don't get lost in the many different descriptions of habitats and their features. If you didn't look for common influences and interactions, keep those ideas in mind as you study the material again. Remember that food webs occur in all environments, and are always based on the producers. In addition, keep in mind that changes in environments are normal, but humans tend to cause such rapid change that many living things cannot adapt, and become extinct.

HINTS

Read these suggestions before you check your answers to the EXERCISES.

For question 3: Think about trophic levels. The first answer is not "consumers."

For question 4: Keep thinking about trophic levels.

For question 9: Taiga is one of the biomes.

For question 12: The key word here is "only."

For question 13:	Eutrophication occurs more readily in still water.
For question 18:	This is looking at things in a general way. It has to do with weather patterns.
For question 20:	Look at Figure 25.13 in the text.
For question 22:	Assume that the soil is still there.
For question 23:	Think about whether the temperature goes up or down.

CONGRATULATIONS . . . YOU HAVE COMPLETED CHAPTER 25 !!!

CHAPTER 26

POPULATION DYNAMICS

CONNECTIONS

The last two chapters have discussed both individuals and populations, and the types of interactions which are common among them. As you have guessed from the title, this chapter will emphasize the factors which influence *populations*. You may remember (from the beginning of the text) that natural selection acts on populations, so watch for references to the effects of natural selection as you study this material: the concepts of how populations change and adapt.

Once again, understanding the role of the environment is crucial to this topic, as it has been in other chapters. Dealing with the environment is a part of a population's "strategy" for long term-survival, so pay attention to the examples of this.

The effects of human activities on the populations of other organisms are described here, but a discussion of the human population itself will come in the next chapter. At that time, however, you will see that these same basic principles of population dynamics are universal.

OBJECTIVES FOR THIS CHAPTER

When you have finished Chapter 26, you should be able to:

1. Discuss why we should be concerned with species extinction.
2. Identify the factors that result in a J-shaped population growth curve.
3. Identify the factors that result in an S-shaped population growth curve.
4. Describe several reasons for population crashes.
5. Explain why most species never achieve their biotic potential.
6. Compare and contrast the reproductive strategies employed by various species.
7. Differentiate between density-dependent and density-independent mechanisms.
8. List several types of abiotic and biotic control mechanisms and state how each regulates population size.
9. Discuss the role of death in the evolutionary development of populations.

CHAPTER 26 OUTLINE

I. Populations, Ethics, and Necessity
II. How Populations Change
    A. Population Growth
III. Controlling Populations Through Reproduction
    A. Tapeworms
    B. Chimpanzees
    C. Birds
IV. Controlling Populations Through Mortality
    A. Abiotic Control and Density-Independence
    B. Biotic Control and Density-Dependence
    C. Predation
    D. Parasitism
    E. Competition
    F. Disease
V. The Advantage of Death

CONCEPTS IN BRIEF

Populations, Ethics, and Necessity

Extinction was discussed in the previous chapter, and it comes up again here. The argument of preserving the genetics of wild populations is emphasized. It is clear that these organisms have been successful in dealing with the rigors of the environment, so perhaps there is some way (now or in the future) that we can benefit by using their successful genes for our crops, livestock, and medicines.

Most wild populations are doing well, as you might expect. The examples (given in the text) of organisms whose populations are growing, are plants and animals which were introduced into a new environment, and are not being effectively controlled by the factors which are discussed later in the chapter.

How Populations Change

If a population of organisms found itself in its ideal environment, with nothing to slow its growth except the number of individuals who were reproducing, it would grow rapidly. A graph of its growth would be a J-shaped curve, in which the *rate* of growth increases over time (an exponential curve, for those of you who are mathematically inclined), as more members of the population are able to reproduce. The maximum reproductive ability of a population is its biotic potential.

Periods of very rapid growth are usually short lived because some feature (or features) of the environment will begin to affect the population and the rate of growth will decrease. The effect of these limitations is called environmental resistance. Most environments have a maximum population size (the carrying capacity) which can be supported for long periods of time. As a population gets close to the carrying capacity of the environment, the environmental resistance will have more of an effect and the size of the population will level off, resulting in a growth pattern shown by the S-shaped curve.

Some populations, however, grow very rapidly (surpassing the carrying capacity) and then decrease very rapidly, or crash. This may be their normal pattern, or it may be due to any of several unusual interactions with their environment. If the environment is damaged, it cannot support as many of those organisms in the future; its carrying capacity for them has been lowered.

Controlling Populations Through Reproduction

Species have developed their own patterns of reproduction (a reproductive strategy), through interactions with their environment (natural selection), in which they attempt to produce as many offspring as possible. These may, of course, be affected by local and immediate conditions that result in good or bad reproductive seasons. This should remind you of reproductive fitness from earlier chapters.

The reproductive strategies of the tapeworm and chimpanzee are very different in their mechanics, but the major difference in focus relates to the amount of energy the parents put into each individual offspring. Birds have developed a strategy to maximize the number of offspring which successfully leave the nest, considering the limitations of the environment and the parents' ability to feed them (to provide energy).

Controlling Populations Through Mortality

Death is an important factor in regulating population size. Individuals may die because of abiotic (nonliving) or biotic (living) components of the environment. Abiotic factors (climate changes and pollution) generally affect populations in a density-independent fashion. The result is that both dense and dispersed populations are affected in the same manner.

Biotic controls are more selective in their effects, with some form of negative feedback operating. Examples of this include: predation, parasitism, competition, and disease.

Predators seldom kill all of the prey population. It would seem like a poor strategy, since they (and their offspring) would have to find a new food source. Parasites follow a similar approach to the individuals which they affect. Humans, however, have not always understood these concepts in relation to their own activities.

Territorial behavior and maintaining hierarchies are two forms of competition. Both of these result in dividing up the resources of the environment, but it is not divided equally among members of the group. Consequently, some individuals have an advantage over others, which should remind you of natural selection. Diseases spread (and survive) more easily in large and crowded populations, and therefore have more of a chance to mutate as well.

The Advantage of Death

This section ponders the question of why organisms die naturally, rather than living indefinitely. The biological aspects of the question, rather than the philosophical ones, are pursued. Once parents have reproduced, their genes have been passed on and any new combinations are present in the offspring (think back to meiosis). Since space is limited on the earth, the two generations will probably end up in competition with one another, unless one dies. If environments change, the genetics of the offspring may be more adaptive than those of the parents. Perhaps it is the reproduction of genes, and not individuals, which is stressed in nature.

KEY TERMS

Be sure to make and use flash cards for all of these terms. Suggestions are found in "To The Student" at the beginning of the Study Guide. Page numbers refer to the text.

biotic	739	carrying capacity	740
abiotic population control	749	crash	741
biotic potential	752	density-dependent	752

density-independent	749	population dynamics	735
environmental resistance	740	reproductive strategy	744
hierarchy	756	S-shaped curve	739
J-shaped curve	739	territoriality	755
parasitism	754		

EXERCISES

Check your understanding of Chapter 26 by completing the following exercises. Answers begin on p. 207.

Use the following graph (of a hypothetical population of animals) to answer Questions #1–7.

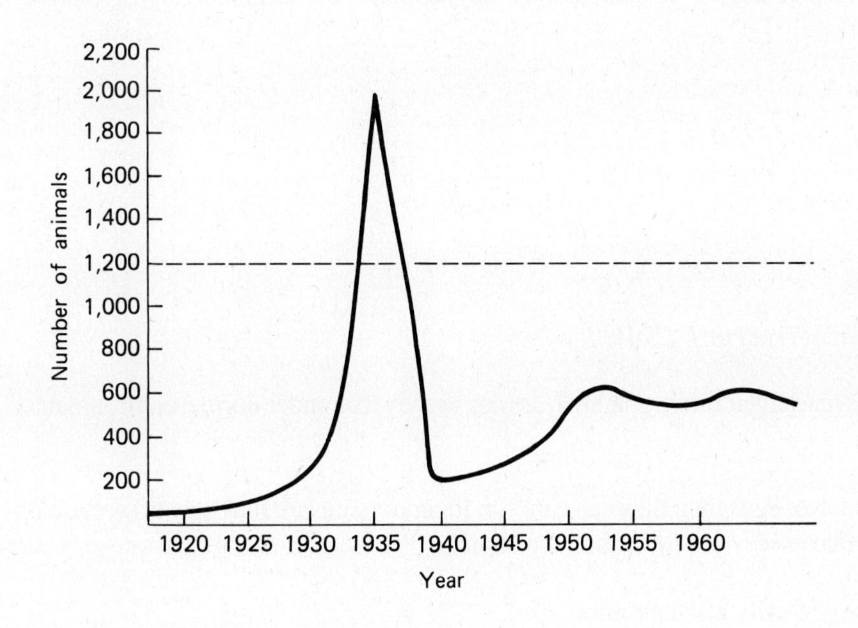

1. When did this population pass the carrying capacity?

2. What time period represents a J-shaped curve?

3. When did the population begin to crash?

4. When did the population begin to recover from the crash?

5. Had this population begun to stabilize by the end of the time period shown on the graph?

6. What may have been responsible for the slow rate of growth before 1925?

7. What may have been responsible for the population's staying below the original carrying capacity after 1955?

True or False Questions

Mark each question either T (True) or F (False).

____8. Most populations reach their biotic potential, as long as they live under normal environmental conditions.

____9. Birds tend to lay fewer eggs than they can care for, in order to ensure that they do not use up all of their available energy in just feeding the babies.

____10. Predators do not generally kill their prey.

____11. Parasites do not generally kill their hosts.

____12. Territorial animals generally share the resources of the environment equally.

Matching Questions

Each question may have more than one answer; write all of the answers that apply. Answers may be used more than once.

____13. biotic components of the environment

____14. weather

a. density-independent factor(s)
b. density-dependent factor(s)
c. neither of the above

____15. diseases

____16. predators

____17. subject to negative feedback controls

Multiple-Choice Questions

Circle the choice that is the best answer for each question.

18. Tapeworms put most of their reproductive energy into
    a. releasing eggs.
    b. caring for their offspring.
    c. searching for a new host.
    d. all of the above receive approximately equal energy.

19. The effects of environmental limits on population growth can be seen in a(n)
    a. J-shaped curve.
    b. S-shaped curve.
    c. both of the above.
    d. none of the above.

20. Current rates of extinction
    a. are reducing the world's gene pool.
    b. are making it easier to produce new crops.
    c. are being reversed by genetic engineering.
    d. all of the above.

21. Density-independent factors in the environment
    a. affect all populations equally.
    b. have a greater affect on small populations.
    c. have a greater affect on dense populations.
    d. affect abiotic populations more severely than they affect biotic populations.

Fill in the Blanks

Complete each statement by writing the correct word or words in every blank.

22. The population size that can survive in an environment for a long period of time is the __________ __________.

23. An S-shaped curve shows a __________ rate of growth than a J-shaped curve. This may be due to __________ __________.

24. One explanation for the necessity of death involves the idea that an animal's body is just its genes' way of making more __________.

Questions For Further Thought

Write your answer for each question in the space provided.

25. Describe some of the reasons why the world is not covered in a layer of houseflies.

SUMMARY

Study the eight items listed in the SUMMARY on p. 760 of the text. You should see that populations are subject to the effects of natural selection, and they change through time. Species which survive today have developed successful strategies for dealing with their environments, however many cannot survive their interactions with the human population and its waste products.

Populations are affected by interactions with their own species as well as with other species. How well they are adapted to the physical characteristics of their surroundings, and whether or not they damage their own environment will affect their long- and short-term survival. Population density, rates of reproduction, and rates of death must all be considered in an evaluation of the current status and possible future of any population.

HINTS

Read these suggestions before you check your answers to the EXERCISES.

For question 1.:	See Figure 26.10 in the text.
For question 2:	See Figure 26.6 in the text.
For question 6:	Assume that this began as a very good environment for these animals.
For question 8:	In order to reach biotic potential, limiting factors cannot be affecting the population.
For question 12:	Think about the "sharing equally" part.
For question 19:	A key word here is "environmental."
For question 22:	See the hint for Question #1.
For question 23:	See Figure 26.9 in the text for the second answer.
For question 25:	See Table 26.1 in the text.

CONGRATULATIONS ... YOU HAVE COMPLETED CHAPTER 26 !!!

CHAPTER 27

HUMAN POPULATIONS

CONNECTIONS

Think about J-shaped population curves (from the previous chapter) as you study the growth of the world's human population in this chapter. The principles of population growth apply to us as well as they apply to the rest of the organisms on the earth. Look for ways in which we may be controlling or eliminating some of the environmental factors that normally limit other populations. What does that do to the carrying capacity of our planet?

You have seen that populations may stabilize or crash. Consider this as you examine the data and predictions for our future populations, and think about how those scenarios apply to us. The next chapter will discuss how our population growth affects the supply of natural resources of the earth.

OBJECTIVES FOR THIS CHAPTER

When you have finished Chapter 27, you should be able to:

1. Briefly outline the nature and growth of human populations through the seventeenth century.
2. List the events that have accounted for major increases in human populations.
3. Discuss the factors which have affected population growth since the seventeenth century.
4. Describe how demographers determine a population's rate of natural increase.
5. Relate crude birth rates and crude death rates to zero population growth and doubling time.
6. Compare and contrast the growth rates among the less developed countries with those in the more developed countries.
7. Describe how age structure pyramids are used to predict population growth.

CHAPTER 27 OUTLINE

I. Early Human Populations
II. The Advent of Agriculture
III. Population Changes From 1600 to 1850

IV. Population Changes After 1850
V. The Human Population Today
VI. The Future of Human Populations
  A. Population Structure

CONCEPTS IN BRIEF

Early Human Populations

In trying to study early human populations, the best we can do is estimate how many people were on the earth at any one time. The estimates are based on the life-style of the people, where they lived, and the carrying capacity of the earth at that time.

It is thought that before 8000 B.C. humans followed a hunter and gatherer life-style, in relatively small bands of people. The overall effect on the environment would have been minimal, as people moved when necessary, and the local environments would have had periodic chances to recover. This, however, changed as improvements in technology resulted in better weapons. Also, the development of language and political systems made hunting groups and food gathering more efficient, resulting in greater environmental impact.

The Advent of Agriculture

Between 8000 B.C. and the mid 1600s human populations were able to increase because of the development of agriculture and tool making. By storing food for future use, more people could survive in the same area, which meant that the carrying capacity of the earth had increased.

Population Changes From 1650 to 1850

From 1650 to 1850 the world experienced continued increase in its carrying capacity for humans. Agricultural practices continued to improve, Europeans began to settle other parts of the world, and attempts were made toward better sanitation. The result was a sudden and serious increase in the size of the human population.

Population Changes After 1850

Rather dramatic population changes have been seen since the 1850s. As always, agricultural practices improved. In addition, the industrial revolution provided better transportation of foods around the world. The big news stories, however, were the acceptance of the fact that bacteria caused disease, and the development of a vaccine for smallpox. As a result, improved medical care lowered death rates around the world. Industrialized nations experienced a drop in birth rates, but the nonindustrialized areas of the world did not. With lower death rates and no lower overall birth rates, the population growth continued.

The Human Population Today

Even though there has been an overall decrease in the *rate* of growth, at the present time the population of the earth is still increasing at an unbelievable rate. Birth control, liberalized abortion laws, and changing attitudes about family size and the role of women in society have contributed to the slowing of the world's growth rate.

Until the world widerate of growth reaches zero, the population will not have a chance to stabilize. Within forty years the population will be twice as large as it is today, and the need for human services and resources will grow with it. The majority of this increase will be in the less developed countries.

The Future of Human Populations

It seems clear that the human population will continue to increase rapidly well into the twenty-first century. Age structure pyramids allow us to predict some of the future growth by knowing what portion of any regional population is in the prereproductive, reproductive, or postreproductive age groups. Many social implications accompany these data as well: from who is of working age and who is not, to transportation, housing, and medical needs.

The carrying capacity of the earth for humans (given our current levels of technology and agriculture) is not known. And although some people are hopeful, we don't know if we will be able to increase it with future advancements. Another very important concern is the question of what quality of life will be available for our doubled population, even if support services and food supplies are doubled along with it, which is not very likely to happen.

KEY TERMS

Be sure to make and use flash cards for all of these terms. Suggestions are found in "To The Student" at the beginning of the Study Guide. Page numbers refer to the text.

age structure pyramid	775	less developed countries (LDCs)	785
crude birth rate	771	more developed countries (MDCs)	785
crude death rate	771	rate of natural increase	771
demographer	775		

EXERCISES

Check your understanding of Chapter 27 by completing the following exercises. Answers begin on p. 207.

1. Draw two age structure pyramids. "A" should represent a population which is stable now, and is likely to remain stable in the future. "B" should represent a population which has been growing rapidly in the past, and is likely to continue to grow rapidly.

A B

2. Use the information in Table 27.1 to calculate the rate of natural increase for the following populations.

   a. Europe

b. United States

c. Africa

d. World

True or False Questions

Mark each question either T (True) or F (False).

_____3. The human population has never gotten close to the earth's carrying capacity for people.

_____4. About 8000 B.C. the population of the earth began to drop steadily until the 1600s.

_____5. Women generally live longer than men.

_____6. Some of the most densely populated areas of the world today are in North America.

_____7. A population can be expected to grow slowly in the future if most of its population is just below the age at which they reproduce.

Matching Questions

Each question may have more than one answer; write all of the answers that apply. Answers may be used more than once.

The following questions deal with population differences between 1970 and 1990.

_____8. rate of natural increase for the world

_____9. rate of natural increase for Latin America

_____10. population of Europe

_____11. doubling time of China

_____12. crude death rate of North America

_____13. crude death rate of the Soviet Union

a. increased
b. remained the same
c. decreased

Multiple-Choice Questions

Circle the choice that is the best answer for each question.

14. The earth's carrying capacity for humans has _____ since the appearance of humans.
    a. decreased periodically
    b. remained approximately the same
    c. increased periodically

15. When a society changes from being an agricultural society to an industrialized one, it generally shows a(n) _____ in the birth rate.
    a. decrease
    b. stabilization
    c. increase

16. Most of the people who have ever lived on earth have
    a. have been hunters and gatherers.
    b. used agriculture as their main source of food.
    c .lived in industrialized societies.
    d. both b and c.

17. The increase in the world's population between 1600 and 1850 was due to increases in the population in
    a. Europe.
    b. China.
    c. Africa.
    d. North America.
    e. all of the above.

Fill in the Blanks

Complete each statement by writing the correct word or words in every blank.

18. Since the 1850s, people have understood that __________ cause disease. At about that same time the first immunizations began with a __________ for __________.

19. Improvements in __________ techniques have resulted in __________ in human populations since humans first appeared.

20. The population of the world will double in approximately __________ years. That means that there will be close to __________ people on Earth then.

21. The part of the world with the highest percentage of its population in the prereproductive age range is __________.

Questions For Further Thought

Write your answer for each question in the space provided.

22. a. Explain why the rate of natural increase for humans must reach zero (or lower) before the world's population can stabilize. Why can't it stabilize with even a small rate of natural increase?

b. Does a zero rate of natural increase mean a zero birth rate?

SUMMARY

Study the seven items listed in the SUMMARY on p. 778 of the text. You have seen that the doubling time for the human population has decreased dramatically in the past three hundred and fifty years, which means that we are growing at faster and faster rates. Advances in agriculture, technology, and medicine have largely been responsible for decreasing the *death* rate worldwide. Any decreases in the *birth* rate have not yet been great enough to result in a stabilization of the earth's population. Consequently we are in a period of extremely rapid growth that is going to continue for many years. Currently, the most rapid increases in population are occurring in the less developed countries, and all predictions indicate that such trends will continue.

HINTS

Read these suggestions before you check your answers to the EXERCISES.

For question 1:	See Figure 27.7 in the text.
For question 2:	See Table 27.1 in the text.
For question 3:	Think about earlier periods of time, not just now.
For question 6:	See Figure 27.8 in the text.
For question 8.:	See Table 27.2 for all of these matching questions.
For question 19:	The second answer is "increases" or "decreases."
For question 21:	See Table 27.2 in the text.

CONGRATULATIONS . . . YOU HAVE COMPLETED CHAPTER 27 !!!

CHAPTER 28

RESOURCES, ENERGY, AND HUMAN LIFE

CONNECTIONS

Chapter Twenty-Seven made it very clear that the human population of earth is growing rapidly. Of course, as the number of people increases so does the need (and demand) for food, medical care, shelter, and many other commodities and services. This chapter examines the impact we have had on the earth's resources, and considers what may happen with them (and to us) in the future.

As you read this material, look for examples of situations in which there are many causes behind an environmental problem. Such things are generally not as simple as the media or politicians would like to have you think. Be prepared to consider your own role in this topic, as a member of the world community. Keep in mind the interconnectedness of everything (from Chapter Twenty-Five) and the many aspects of behavior (Chapter Twenty-Four) which come to play in these situations. The next chapter will examine the environmental impacts of our resource use in more detail.

OBJECTIVES FOR THIS CHAPTER

When you have finished Chapter 28, you should be able to:

1. Discuss the major food resource problems facing human populations today.
2. List some possible ways the world's food supply may be increased.
3. Relate the food pyramid to the world's food problem.
4. Compare the dietary efficiencies of eating different types of animals.
5. Describe some reasons for severe water shortages.
6. Give examples of nonrenewable resources.
7. Explain why recycling is an energy-efficient and environmentally sound process.
8. Name some materials that are currently recycled.
9. Describe several energy sources, citing the advantages and disadvantages of each.
10. Explain the role of conservation in solving energy problems.

CHAPTER 28 OUTLINE

I. Renewable Resources: Focus on Food and Water
  A. Food and the Present Crisis
    1. The Global Implications of Food Resources
    2. Some Global Realities
    3. Projections of Food Supply
    4. Farming the Earth's Jungles
    5. Domestic Animals as Food
    6. Fishing
    7. Solutions to the Hunger Problem
II. Water and the Coming Crisis
III. Nonrenewable Resources
IV. Recycling - And Around It Goes
V. Energy
  A. Energy fom Fossil Fuels
  B. The Geopolitics of Fossil Fuel
  C. Energy from Water
  D. Energy from the Wind
  E. Energy from the Earth
  F. Energy from the Sun
    1. Active Solar Systems
    2. Passive Solar Systems
    3. Photovoltaics
  G. Energy from the Atom
VI. Encouraging Conservation and Increasing Efficiency

CONCEPTS IN BRIEF

Renewable Resources: Focus on Food and Water

Since food is a renewable resource, there is a theoretical chance that the needed amounts could be produced in the future. This assumes increased yields, new food sources (which people typically refuse to eat anyway), favorable weather, and no major attacks on crops by new pests or predators. Barely enough food is produced now, even though it doesn't get to everyone. As the population continues to increase, the carrying capacity of the earth will have to be increased as well.

Like many other resources, **food supplies** are not evenly distributed, so the problems of transportation, economics, waste, losses to normal pests, international politics, and civil wars must also be overcome before the food gets to everyone who will need it. The projections are, that even if all of the above happens, the quality of life may still go down due to the total combination of demands by our growing population.

The effects of **starvation** (including being undernourished and malnourished) can be seen throughout the world. Although adults may recover if rescue efforts arrive in time, children often suffer permanent damage due to poor brain and skeletal development. This, of course, has many social implications as these children age.

Additional food harvests from the **jungles** and **oceans** seem like good ideas, but are not biologically sound ones given what we currently know. Farming the jungles requires that we develop methods suitable to the area, instead of relying on the conventional methods which are so effective in the temperate regions of the earth. The seas are commonly overharvested now. Our lack of understanding of the population dynamics of many marine organisms makes it difficult for us to plan for effective future yields.

Eating low on the **food chain** results in more efficient use of energy (and water). Animal protein, however, is a convenient source of several amino acids (which can also be acquired from an appropriate selection of plants), so selecting efficient livestock for an area is important. Poultry produce more meat per pound of plant tissue (that they eat) than do cattle, sheep, goats, or pigs.

Water and the Coming Crisis

Fresh water is needed for drinking, food preparation, agriculture, and industrial uses, yet it only makes up three percent of the earth's total water supply. Many areas of the United States are using their underground water faster than the supply can be replaced. Deeper wells are only a temporary solution. Evaporation and pollution are the main reasons that some water supplies are lost to us. If the pollution levels are high enough, the water may even become a nonrenewable resource.

Nonrenewable Resources

Metals, ores, coal, and natural gas are considered to be nonrenewable; the supply is fixed. As with renewable resources, they are not evenly distributed among the regions of the world. Current technologies have not allowed us to locate all of the deposits, nor estimate their size. Long-term plans must accept the fact that the ability to find new sources is limited.

Recycling - And Around It Goes

Recycling allows us to reuse the nonrenewable resources mentioned above, along with paper, oil, and plastic products. This will decrease the need to find and process new supplies. In addition, much less energy is used in manufacturing new products if they are made from recycled materials. Unfortunately, American consumers use disposable products, throw away many recyclable goods (unless the money they receive for it is high enough), and political lobbies fight legislation to require deposits (which would encourage recycling).

Energy

In Chapter Five you studied energy and the laws of thermodynamics. Those principles apply here as well. Figure 28.16 in the text shows you that energy usage per person has increased dramatically through time. Even at that, it appears that we do have sufficient supplies. The issue to consider is which sources to use in various situations and the serious environmental cost (both long and short term) of using each type.

Coal, oil, and **natural gas** supplies are effectively nonrenewable. Burning any of these creates serious air pollution problems (which will be considered in more detail in the next chapter), including an aggravation of the greenhouse effect. The countries of OPEC control most of the world's supplies of oil, which makes the entire issue very political as well.

Moving **water, wind,** and **sunlight** provide available energy in a renewable fashion. Of course using each of these also has its environmental consequences. Efficient use of water and wind energy is limited, at least on a large scale, to certain parts of the world. Solar energy is widely available (and free), but (except for President Carter's efforts) there has not been sufficient research money allocated for encouraging its use. Systems which use energy from the sun for heating can be added to existing structures (as active solar systems), or put into the design of new structures (either active or passive systems). Photovoltaic systems for converting sunlight into electricity are effective but generally expensive and require large amounts of space.

Geothermal energy locations are localized, limited, and nonrenewable. **Nuclear** energy (through fission) is used for producing electricity, but poses very serious long-term environmental problems.

Encouraging Conservation and Increasing Efficiency

Using less energy, and wasting less of what we do use, are very good ways to conserve and reduce the environmental damage (and possibly political upheaval) associated with resource and energy usage. This would certainly require a change in the outlook (and behavioral patterns) of many Americans. Conservation, recycling, and research will have to be funded and encouraged.

KEY TERMS

Be sure to make and use flash cards for all of these terms. Suggestions are found in "To The Student" at the beginning of the Study Guide. Page numbers refer to the text.

active solar systems	808	nonrenewable resources	799
fossil fuels	802	passive solar systems	808
geothermal energy	807	recycling	800
hydroelectric energy	806	renewable resources	781
kwashiorkor	783	solar energy	807
malnourished	784	undernourished	784
marasmus	783		

EXERCISES

Check your understanding of Chapter 28 by completing the following exercises. Answers begin on p. 207.

Use Figure A (of Table 28.2 in the text) to answer Questions #1–3.

1. What was the source of the herbivores' calories?

2. What percentage of the energy in the herbivore level is passed to the primary carnivores?

3. What happened to the rest of the calories that were in the herbivore level?

True or False Questions

Mark each question either T (True) or F (False).

____4. Within the next five years most of the people of the world will live in more developed countries.

____5. Children generally recover fully from the effects of kwashiorkor after they begin to have an adequate diet again.

____6. Enough food is produced in the world to feed the people that are here now.

____7. Jungle soils are generally rich in the nutrients needed for growing crops.

____8. Relying on a small variety of food crops makes it easier to keep them protected from new types of pests.

____9. It takes less energy to manufacture cans from recycled materials than from newly mined ore.

____10. Political problems are often associated with areas experiencing famines.

____11. Although nonrenewable resources are not distributed evenly around the world, most renewable resources are distributed evenly.

Matching Questions

Each question may have more than one answer; write all of the answers that apply. Answers may be used more than once.

____12. coal

____13. forests

____14. solar energy

____15. aluminum

____16. food

a. a renewable resource
b. a nonrenewable resource

Multiple-Choice Questions

Circle the choice that is the best answer for each question.

17. Marasmus is caused by a diet low in
   a. calories.
   b. protein.
   c. both a and b.

18. One of the effects of the famine in Africa is that the carrying capacity of the area has
    a. been increased due to the relief efforts.
    b. stayed approximately the same because the number of cattle has been lowered.
    c. decreased because the land has been damaged by farming and grazing.

19. Which of the following is generally the most efficient and least costly to raise?
    a. cattle
    b. chicken
    c. goats
    d. sheep

Fill in the Blanks

Complete each statement by writing the correct word or words in every blank.

20. When a person no longer has food to eat, their body first uses stored __________ for energy. After that they begin using stored __________.

21. Most of the plant protein that people eat comes from __________, __________, __________, and __________, even though there are over __________ species of known edible plants.

22. The two main problems associated with losing available fresh water are __________ and __________.

23. Photovoltaic systems convert __________ energy into __________.

24. Using less energy (__________) and using materials again (__________) will help the world's resources last __________.

Questions For Further Thought

Write your answer for each question in the space provided.

25. Explain how the law of the commons may result in overgrazing when several people have cattle on public lands.

SUMMARY

Study the thirteen items listed in the SUMMARY on p. 813 of the text. You have seen that many of the earth's resources occur in limited amounts, and are being used by more people every day. Even if a resource is renewable it is necessary to examine the environmental consequences of its use. Long range planning for all of these issues is very important, and often not done effectively. It is common for people to view themselves as being separate from these effects rather than being part of the world environment.

Because of the uneven distribution of resources, it is not always possible for any one region of the world to acquire and use a variety of materials and energy supplies. International politics and economics play an important role in much of the decision making concerning resource use. Increased conservation and recycling efforts can have a serious positive impact and our overall quality of life.

HINTS

Read these suggestions before you check your answers to the EXERCISES.

For question 1: There are 4846 Kilo-calories (per square meter of area per year) available in the herbivore level (add 3368 + 1478), and 450 in the primary carnivores.

For question 3: Don't forget the laws of thermodynamics.

For question 6: This asks about production, not distribution.

For question 7: The key word here is "soils."

For question 8: This is asking about the effects of monocultures.

For question 18: Think about how carrying capacity is affected, as described in the previous two chapters.

For question 21: The first four answers can be written in any order.

CONGRATULATIONS . . . YOU HAVE COMPLETED CHAPTER 28 !!!

CHAPTER 29

BIOETHICS, TECHNOLOGY AND ENVIRONMENT

CONNECTIONS

During the past several chapters you have been presented with issues of behavior, population growth, and the effects of humans on the environment. That theme continues here as you study a variety of pollutants (which are found throughout the world) and their effects. Look for the sources and actions of these agents. Notice both their long- and short-term consequences to various aspects of our environment.

At this point in the course you have sufficient biological knowledge to ask questions in a scientific fashion and to examine and analyze data in order to arrive at an answer. That is exactly what this chapter asks you to do on a daily basis. You are reminded that how we treat the world and the other organisms in it is entirely up to us, and is based on the decisions we all make every day. Keep that in mind as you study this final chapter.

OBJECTIVES FOR THIS CHAPTER

When you have finished Chapter 29, you should be able to:

1. Describe some issues addressed by bioethics.
2. Define *pollutant* and list four ways in which pollutants are harmful.
3. Identify the major air pollutants, and list the sources and effects of each.
4. Identify the major water pollutants, and list the sources and effects of each.
5. Describe environmental problems associated with pesticide use.
6. List some sources of radiation and describe the risks associated with nuclear power.
7. Discuss ways in which people make biological decisions in their daily lives.

CHAPTER 29 OUTLINE

I. The Ethics of Doormats
II. Environmental Pollution
  A. Air Pollution
    1. Carbon Monoxide
    2. Nitrogen Oxides
    3. Sulfur Oxides
    4. Hydrocarbons
    5. Particulate Matter
  B. Water Pollution
    1. Sewage
    2. Chemical Pollution of Water
    3. Heat Pollution of Water
  C. Pesticides
  D. Radiation
    1. Pollution from Nuclear Reactors
III. Hidden Decisions
IV. The Future

CONCEPTS IN BRIEF

The Ethics of Doormats

Questions of bioethics are not always easy to answer, even if you think that you have a firm stance on environmental issues. This section suggests the need to focus on our values while we consider the long range effects of our role in, and interactions with, the ecosystem of the earth.

Environmental Pollution

Pollutants are not new, but people make and release new kinds every day. The focus here is on air pollution, water pollution, pesticides, and radiation. All of these are present in your own environment now (regardless of where you are), and may be causing short- and long-term damage to you and your surroundings. Pollution caused by natural events (volcanoes, fires, etc.) is often beyond our control, so human-caused problems are emphasized in this chapter.

Most **air pollution** is caused by burning materials, such as the fossil fuels used in vehicles, homes, and industry. Automobiles are the major source. Table 29.1 in the text summarizes the effects of four important air pollutants. Notice that they affect your immediate health, as well as causing chronic heart and respiratory problems. These effects are in addition to their role in corroding metals, stone, and many other materials. Cigarettes, of course, are a source of air pollution.

People have always dumped their wastes in nearby bodies of **water**, but now there are too many people and too many exotic chemicals for the natural aquatic environment to handle it all. Sewage treatment is designed to remove organic materials and large particles, which can be done very effectively. Chemical pollutants are generally not removed, which results in a strong possibility of polluted drinking water downstream. Heating an aquatic environment is likely to cause a change the species composition, since many aquatic organisms are adapted to narrow temperature ranges and cannot tolerate warmer water.

Pesticides are designed to kill living things, which can be good and bad. If you want to kill local pests, the effects may be considered good. However, the pesticide may spread through the environment, and is likely to kill many other species as well. Long-lived pesticides may stay potent for many years. Some may remain in living tissues, and then are passed along food chains, concentrating in the predators at higher trophic levels.

Background radiation has been present throughout the evolution of life on earth. The current concern is the concentrated **radioactive materials** (with very long half lives) which are associated with medical radiation, nuclear testing, and nuclear reactors. Essay 29.5 in the text describes the process by which radioisotopes can generate free radicals in your cells, and perhaps alter your DNA.

Safely storing the fissionable materials and waste products associated with the nuclear energy industry has presented a serious problem, with no solution yet that is acceptable to the world community. The text presents sobering (and true) examples of the dangers of nuclear reactor accidents.

Hidden Decisions

You have read several parts of the text (in this chapter and previous chapters) which stress the interconnectedness and interdependence of all parts of the world ecosystem. You are reminded of that, once again, in this section. What you may normally think of as simply personal, political, academic, or business decisions are often of great importance to the biological well-being of the earth as well. You are encouraged to become more aware of the environmental impact of your choices.

The Future

Yes, this really is the final section of the last chapter! It is now your task to apply what you have learned and make deliberate choices about the future, using the abundant data that are available to you. Your generation and the next will have to deal with the results.

KEY TERMS

Be sure to make and use flash cards for all of these terms. Suggestions are found in "To The Student" at the beginning of the Study Guide. Page numbers refer to the text.

Term	Page	Term	Page
bioethics	817	pesticide	828
carbon monoxide	821	photochemical smog	823
chlorofluorocarbons (CFCs)	822	pollutant	
hydrocarbon	824	radiation	829
nitrogen oxide	823	sewage sludge	825
nuclear winter	836	smog	823
ozone	822	sulfur oxide	823
particulate matter	824	temperature inversion	821
pest			

EXERCISES

Check your understanding of Chapter 29 by completing the following exercises. Answers begin on p. 207.

1. Arrange the following events of sewage treatment in sequence from first to last.
   a. sludge forms
   b. chlorine is added

c. fluids pass through a screen filter
d. air is pumped through the fluids
e. bacterial breakdown of waste is speeded up
f. pollutants are removed if the facility is equipped properly

____ ____ ____ ____ ____ ____

First Last

True or False Questions

Mark each question either T (True) or F (False).

____2. Bioethical questions are usually easy to answer if you follow a valid process of scientific reasoning.

____3. You make biological decisions every day.

____4. A thin layer of ozone covers the entire earth.

____5. A meltdown of a nuclear reactor is theoretically possible, but safeguards in power plants have prevented it from ever happening.

____6. Newer chemical pesticides are designed to break down into harmless materials.

____7. Some pollutants may act together in the environment to produce unexpected effects.

Matching Questions

Each question may have more than one answer; write all of the answers that apply. Answers may be used more than once.

____8. sulfur oxides

____9. carbon monoxide

____10. particulate matter

____11. nitrogen oxides

a. irritates lungs
b. affects plant photosynthesis
c. damages buildings
d. reduces the amount of oxygen in your blood
e. all of the above

Multiple-Choice Questions

Circle the choice that is the best answer for each question.

12. Atmospheric studies show that air pollutants
    a. are limited to major cities around the world.
    b. occur throughout the Unites States, but many other areas are free of them.
    c. occur throughout most of the world, except for a few remote areas.
    d. occur around the entire earth.

13. Temperature inversions occur when
    a. pollution levels go above acceptable limits.
    b. rain fails to fall for several days and results in higher levels of air pollution.
    c. a layer of cooler air is trapped beneath a layer of warmer air and the air pollution cannot disperse.
    d. several warm sunny days in a row heat the ground temperature higher than that of the air above it.

14. Radioactive wastes can be decontaminated by
    a. burying them in the ground and allowing the interaction with the soil to inactivate the radioisotopes.
    b. dropping them in deep ocean trenches and allowing the water pressure to compact them and hold them in place.
    c. waiting until at least twenty half lives of the material have passed.
    d. both a and b, but not c.
    e. all of the above.

15. Most chemical pesticides that are intended for insects have been shown to
    a. kill all insects.
    b. prevent insects from developing resistance to their effects.
    c. both a and b.
    d. none of the above.

Fill in the Blanks

16. The greatest amount of air pollution in the United States comes from __________.

17. The layer of __________ in the atmosphere filters the sun's harmful __________ rays, therefore keeping rates of skin cancer down.

18. High levels of __________ __________ are common in traffic jams, and may affect your ability to drive. This pollutant is also a major compound of __________ smoke.

19. Acid rain commonly contains __________ and __________ acids.

20. Heat pollution can result in __________ oxygen being present in a body of water.

21. Burning coal is the major source of __________ __________, which is an air pollutant.

22. Some pollutants do not show any effects in the environment until they build up to a certain level. This is known as the __________ effect.

Questions For Further Thought

Write your answer for each question in the space provided.

23. Explain why freshwater algae contain low levels of DDT (rather than having none), and why fish-eating birds have high levels.

SUMMARY

Study the eight items listed in the SUMMARY on p. 838 of the text. You have seen that humans are an integral part of the world ecosystem. Our rapid technological advances have produced thousands of new chemicals which are now free in the air, soil, and water. The rate at which we have released these (as well as the massive quantities) has made it difficult for many species to adapt, and has also created numerous health problems for ourselves. Many of the long-term results of our actions are not yet known.

It is certainly time for everyone to clarify their personal values and to act accordingly toward the environment. There are numerous ecological questions which do not have clear answers, but now that you have training in biology you are well equipped to seek the solutions.

HINTS

Read these suggestions before you check your answers to the EXERCISES.

For question 4:	Think about the effects of chlorofluorocarbons.
For question :5	Did you read the section about Chernobyl?
For question 10:	Be sure to read Table 29.1 *and* the text.
For question 13:	This question asks about the *cause* of temperature inversions, not the results.
For question 14:	This is asking what works, not just what has been tried.
For question 16:	Did you drive to class by yourself today?
For question 20:	The answer is either "more" or "less."
For question 21:	Don't answer "air pollution" here.

CONGRATULATIONS . . . YOU HAVE COMPLETED CHAPTER 29 !!!

ANSWERS TO CHAPTER EXERCISES

If you have any incorrect answers, reread the* HINTS *section of the Study Guide for that chapter. Also look for help and answers in your text.

CHAPTER 1

1. E,D,A,C,F,B	2. T	3. F	4. F	5. T	6. F
7. E 8. A,B,D	9. E	10. A	11. D	12. D	13. C
14. B 15. A	16. A	17. D	18. A	19. A	20. A

21. increased 22. use or need, naturally, inherited
23. Galapagos, Ecuador, finches 24. survive, reproduce
25. In artificial selection, humans (not natural events) choose the parents, parents are chosen for desired (not survival or reproductive) characteristics, and change usually happens faster.
26. His father felt that Charles would disgrace the family, and that such a voyage was not appropriate for someone who intended to be a minister.
27. a. He believed in special creation and the permanence of species.
    b. His observations confirmed what Lyell had written, and he recorded natural variation around the world. The effects of the environment on animals in the Galapagos made a great impact on him. Finally, the writings of Malthus fit well with what he knew about artificial selection.

CHAPTER 2

1. C,E,D,F,B,A	2. F	3. T	4. T	5. F	6. T
7. D 8. B	9. C	10. A	11. D	12. A	13. B

14. D 15. inductive, hypothesis, testable 16. cells
17. movement 18. evolve, adapt, respond
19. The results of the control are compared to the results of the rest of the experiment. This allows you to see if there is any effect from the one thing that was different between the two parts of the experiment.
20. See the explanations on pp. 39 & 40 for the characteristics that you picked.

CHAPTER 3

1. See Figure 3.7 on p. 58.

2. See Figure 3.10 on p. 61. The covalent bonds are represented by the spots where the outer electron shells of oxygen and hydrogen overlap.

3. T	4. T	5. F	6. F	7. A,B	8. A,C,D
9. A,C,D	10. A,B,C		11. C	12. B	13. C
14. A	15. B	16. covalent, hydrogen			

17. electrons, 8, electrons
18. polymer, nucleotides, RNA & DNA
19. These letters are the symbols for the elements carbon, hydrogen, nitrogen, oxygen, phosphorus, and sulfur. Together these make up about 99 percent of the material found in all living things on Earth.
20. The shape of the active site on the enzyme fits the shape of the substrate. When they join together the enzyme changes shape and the fit is even better.

CHAPTER 4

1.

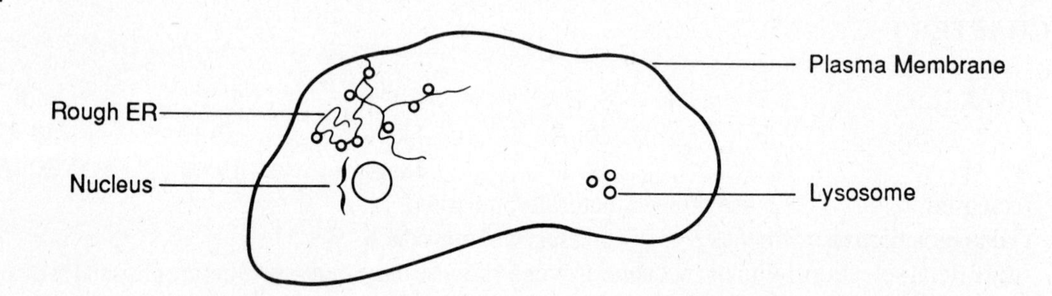

Functions:

Plasma membrane separates the cell from the environment, regulates the cell's interior, determines the cell type, attaches to the cytoskeleton.
Nucleus houses DNA, regulates cell metabolism and reproduction.
Rough ER manufactures proteins.
Lysosome stores digestive enzymes.

2. a. osmosis b. phagocytosis c. diffusion d. exocytosis e. pinocytosis

3. T		4. F	5. F	6. T	7. T	8. D
9. B,C,E		10. A,B,C		11. C,E	12. H	13. A,B,D,F
14. E	15. C		16. E	17. A	18. C	19. B
20. cells		21. osmosis, cell wall				

22. phospholipid, protein, carbohydrate
23. more. More active cells require greater amounts of energy, most of which is provided by reactions that take place in association with mitochondria.
24. Electron microscopes give higher magnification and greater resolution than light microscopes. These allow scientists to study cellular details that were not visible before.

CHAPTER 5

1. B,A,E,D,F,C 2. C,E,F,A,D,B

3. a. stroma b. photosystem I and II c. membranes d. thylakoid
   e. glucose formation f. accumulating protons

4. F	5. T	6. T	7. T	8. B	9. C
10. C	11. A	12. D	13. A	14. C	15. A
16. B,C,D	17. A	18. A	19. D	20. A	

21. membrane, thylakoid
22. mitochondrion
23. ATP, hydrogen, membrane
24. 4
25. 36, 2
26 The light-independent reactions of photosynthesis, which produce glucose, cannot run without the products of the light-dependent reactions (which cannot run in the dark). Fertilizer would provide needed materials for growth, but without the formation of new glucose the plant will starve to death.

CHAPTER 6

1. a. 2 b. 4 c. 2
2. a. meiosis b. 3,1,4,2
3. a. AUGCCGAUUGCAUAA b. MET-PRO-ILE-ALA-Stop
   c. ATGCCGATTGCATAA

4. F	5. F	6. T	7. F	8. T	9. C
10. B	11. A	12. A	13. C	14. B	15. A
16. C	17. C	18. B	19. A	20. A	21. B

22. meiosis, 4, haploid, 16
23. transcription, translation
24. 3, bases
25. In order for a protein to function properly, its amino acids must be in a specific sequence. That sequence is determined by the sequence of mRNA codons, which is determined by the sequence of bases on the DNA molecule.
26. Only certain *sections* of the DNA molecule are read in each type of cell. Therefore, only certain kinds of proteins are manufactured in each type of cell. The cell's type is determined by the types of proteins it produces.

CHAPTER 7

1. a. a and g b. d c. No. Crossing over occurs between similar regions of paired homologous chromosomes, not between two regions of one chromosome.

2. F	3. F	4. F	5. F	6. F	7. F
8. E	9. B	10. A	11. A,C	12. A	13. C,D
14. A,B,C,D	15. A	16. D	17. B	18. B	19. B

20. test, yellow, Gg, yellow podded
21. evolution
22. phenotype
23. multiple, A,B,O
24. incomplete dominance, multiple alleles, epistasis, polygenic inheritance, environmental conditions, age, sex
25. Natural conditions don't always allow for the required random mating, large populations, and isolated populations that are necessary for Hardy-Weinberg to be effective.

CHAPTER 8

1. This could include: making crops resistant to various pests, poor growing conditions, and herbicides; increasing the size of edible parts and yields; increasing nutritional content.

2. T 3. T 4. T 5. T 6. A, possibly D
7. E 8. D, possibly A 9. B 10. A,C, possibly D
11. C 12. A 13. C 14. C 15. B

16. restriction, plasmid, ligase, recombinant 17. clone
18. This would prevent the medical problem from occurring at all, rather than just treating the symptoms.
19. It prevents the release of organisms which are harmful or which could mutate and become harmful.

CHAPTER 9

1. F,D,E,C,A,B 2. F 3. T 4. F 5. T
6. T 7. B,C,D,E 8. E 9. C 10. A,B,C,D,E
11. C 12. B,D,E 13. A 14. D 15. C
16. D 17. C 18. A 19. B 20. B

21. free oxygen
22. Miller, atmosphere, electricity, amino acids 23. life force
24. microspheres, coacervates, liposomes
25. Free oxygen in the beginning would have prevented the formation of many important compounds. Later, free oxygen formed the ozone layer which resulted in a safer environment for life on land.
26. Daughters of protobionts which were successful in *their* environment would also be more likely to be successful if they had a similar chemistry.

CHAPTER 10

1. Your early population graph should look like the top one in Figure 10.8 B, and your current population should be like the bottom one in Figure 10.8 B.

2. F 3. T 4. F 5. T 6. T 7. F
8. B 9. A,B 10. A,B 11. C 12. A 13. B
14. A 15. A 16. C 17. A 18. D 19. C

20. Hardy-Weinberg, alleles 21. Mendelian genetics
22. small, bottleneck, founder
23. Cretaceous, earth, asteroid, dust, photosynthesis
24. These variations are not mutations. They do not affect the gene pool and will not be inherited by the offspring.
25. Having multiple sets of chromosomes is usually fatal in animals. Animal hybrids that do survive are generally infertile or may not be able to compete well with the members of the parent populations.

CHAPTER 11

1. Kingdom, Phylum or Division, Class, Order, Family, Genus, Species 2. a. yes b. Fungi
3. F 4. T

5. T	6. F	7. T	8. B	9. D	10. B,C
11. A	12. A	13. B	14. F	15. D,E,F	
16. E	17. A	18. B	19. A	20. E	

21. hyphae, mycelium, fruiting body

22. Actinomycetes, antibiotics 23. Monera 24. Sporozoa

25. Protista, energy

26. *Euglena* shows characteristics that are usually found only in protists, or plants, or animals.

CHAPTER 12

1. ACROSS: a. frond b. algae c. Gingko d. duckweed e. Ephedra f. root DOWN: a. flower

2. T	3. F	4. F	5. F	6. B	7. C
8. A	9. E	10. B	11. A,B,D		12. E
13. A,B,D		14. A	15. B	16. B	

17. phycoerythrin, red, deep 18. chlorophyll a, chlorophyll b

19. 100, Rhodophyta, brown 20. increase 21. starch

22. dry

23. Although you really can't tell by looking, the moss would not have true leaves, stems, or roots. It *would* have a visible sporophyte (rather than cones) and would probably be smaller than the conifer.

CHAPTER 13

1. a. brain b. digestive tract c. coelom d. nerve cord e. vertebrae f. anus/cloacal opening g. mouth h. gill slits

2. Animal, Chordata, Mammalia, Primates 3. T

4. T	5. F	6. F	7. T	8. C,D,E,F,G,H,I	
9. D,E,F,G,I		10. E,F,G		11. J	12. E,F,G,H,I
13. B,C,D,E,F,G,H,I		14. B	15. C	16. B	17. D
18. mouth		19. bilateral		20. cnidarians	

21. soft, mantle

22. Amphibians faced the problems of dry air, large changes in temperature, and no water for buoyancy. The advantages were that there were fewer competitors, fewer predators, abundant food, and a rich oxygen supply.

CHAPTER 14

1. a. stigma b. style c. carpel d. ovary e. anther f. filament g. stamen h. petal i. sepal j. ovule

2. a. E b. A c. J d. B e. D f. J g. E,J

3. T	4. T	5. F	6. T	7. F	8. F
9. D,E	10. B	11. A	12. F	13. A	14. C
15. B	16. B	17. A	18. pollen grain		
19. root, stem, longer		20. longer			
21. gametangia, sexual		22. auxin			

23. a. An environment which remains stable over many generations.
    b. Asexual reproduction passes the same (and in this case successful) genotype to all of the offspring, which should suit them to the same environmental conditions.
24. Both pollen and seeds offer protection (during the reproductive cycle) from dry air and other harsh environmental conditions found on land.

CHAPTER 15

1. a. ovary b. oviduct c. uterus d. bladder e. urethra f. clitoris g. labia minor h. labia major i. cervix j. vagina k. rectum l. anus m. prostate n. urethra o. penis p. glans penis q. vas deferens r. bladder s. seminal vesicle t. Cowper's gland u. rectum v. anus w. epididymis x. scrotum y. seminiferous tubules z. testis aa. scrotum

2. T	3. F	4. F	5. F	6. F	7. F
8. Y,Z	9. A	10. I,J	11. W	12. B	13. C
14. M,S,T,Y	15. O	16. Q,AA	17. I	18. O,P	19. C
20. A	21. B	22. C	23. C		

24. external, eggs, environment 25. estrogen, uterus, fertilized
26. acidic, alkaline
27. You might include features such as: 100 percent effective, easy to administer, long lasting, easily and quickly reversible, no harmful side effects, effective for males *and* females, inexpensive (or free), socially acceptable for everyone.

CHAPTER 16

1. a. morula, gastrula, 1st cleavage, neural stage, blastula, 2nd cleavage b. 3,5,1,6,4,2

2. T	3. F	4. F	5. T	6. F	7. A
8. C	9. H,I	10. E	11. A	12. F	13. C
14. C	15. A	16. A	17. C	18. C	19. B
20. B	21. A	22. E	23. 266	24. genes	

25. germ, ectoderm, mesoderm, endoderm
26. cleavage, oviduct 27. first
28. The genes that control growth of an arm must be identified, and the means of turning the genes on must be understood.

CHAPTER 17

1. a. muscle #2 b. muscle #2 must relax, and muscle #1 must contract c. muscle #4 d. muscle #4 must relax, and muscle #3 must contract

2. F	3. T	4. T	5. F	6. F	7. T
8. T	9. B	10. B	11. B	12. E,H 13. A	
14. J	15. J	16. B,C	17. D	18. F	19. A,E,H
20. B	21. A	22. D	23. C	24. ATP	25. ligaments

26. Haversian canal 27. axial 28. immobile
29. Calcium constantly needs to be replaced in bone throughout your life. Calcium is also required for muscle contraction.

CHAPTER 18

1. C 2. A,B 3. A,B,C 4. B 5. A,B,C 6. A,B
7. T 8. F 9. F 10. T 11. C 12. A
13. D,F 14. B 15. C 16. B 17. A 18. A
19. B 20. thermostat, fever, bacteria, virus 21. nitrogen
22. protists, animals 23. filtration, reabsorption, secretion
24. The body core temperature is higher than that of the skin, therefore a thermometer in the fist will register too cool a temperature.
25. There are many reasons, but related to this chapter is the fact that worms lose large amounts of water during waste secretion, and they cannot replace it readily in dry dirt.

CHAPTER 19

1. a. amphibian b. bird/mammal c. fish d. reptile
2. See Figure 19.15 in the text. 3. A,D 4. A,B,D
5. T 6. F 7. T 8. T 9. F 10. C,D
11. B,E 12. F 13. A 14. B,C,D 15. A,B
16. C 17. D 18.C 19.E 20. diffusion
21. quit 22. diaphragm, ribs 23. atria, ventricles
24. lacteals, villi
25. The surface area has been increased by the projecting villi and microvilli. This has allowed a more compact body, and has increased the efficiency of absorption.

CHAPTER 20

1. E,B,F,C,A,D 2. T 3. F 4. T 5. T
6. F 7. A 8. C 9. A 10. A 11. B
12. A 13. D 14. B 15. D 16. D
17. plasma, B-, interleukin 18. acidic, alkaline (basic)
19. cells, viruses 20. helper T-, suppresses
21. The vaccination will trigger both a nonspecific response and a primary specific response, therefore you see symptoms. There is no disease because the vaccine will not contain live material (or if it does there is not enough to cause the disease). Memory B-cells are produced which will trigger a secondary response if the agent is ever encountered again, therefore you are immune.

CHAPTER 21

1. A 2. F 3. B,H 4. E 5. C 6. I
7. D,I 8. T 9. F 10. F 11. T 12. F
13. B 14. C 15. A 16. A 17. D 18. B,E
19. A 20. B 21. C 22. B 23. D 24. neuron
25. endocrine 26. dendrite
27. neurotransmitter, axon, dendrite 28. ATP
29. First of all only the dendrite ends are stimulated by the environment, and when they are, they will

pass an impulse on to the axon of that neuron. In addition, at the synapse, only the axon end secretes neurotransmitter substance and only the dendrite end is sensitive to it and able to be stimulated by it.

CHAPTER 22

1. a. cat b. reptile c. fish d. bird   2. cerebrum
3. D   4. C   5. C   6. A   7. T   8. F
9. F   10. F   11. F   12. F   13. F   14. D
15. A   16. B,E   17. A   18. A   19. D   20. C
21. C   22. A   23. forebrain, midbrain, hindbrain
24. reticular   25. stimulant, depressant
26. sympathetic, parasympathetic
27. A reflex arc provides more rapid and consistent responses to stimuli from the environment than would be possible if the animal had to "decide" what to do in every situation. The result is likely to be less risk and damage, consequently a greater chance of survival.

CHAPTER 23

1. C,F,A,D,E,B   2. F   3. F   4. T   5. T
6. A,D   7. D   8. C   9. B   10. C   11. C
12. D   13. C   14. receptors   15. sweet, bitter
16. warm, cold, pain   17. proprioceptors
18. The cones are used for color vision, however they require a certain amount of light. As the light dims they are no longer stimulated and consequently you no longer see color.
19. Moths which are sensitive to volume have a better chance of surviving attacks by bats, whereas those not sensitive to this are more likely to have been killed. As far as we can tell there is no benefit to the detection of different frequencies in these moths, and the ability has not been established.

CHAPTER 24

1. E,B,A,D,C   2. F   3. F   4. F   5. F
6. F   7. B   8. A,B   9. A,B   10. B   11. B
12. B   13. A   14. C   15. C   16. A   17.C
18. B   19. conditioning, classical, habituation
20. competition   21. habituation   22. escapes
23. a. The eggs were hatched in an incubator, and he was the first living thing they saw. They imprinted on him and treated him like their "mother."
    b. The answer to this is not clear, but it seems that imprinted behaviors are not forgotten, and are not really affected by later learning.

CHAPTER 25

1. D,A,C,E,B   2. producers

3. zooplankton, consumers 4. trophic, quaternary consumers
5. T 6. F 7. T 8. F 9. T 10. F
11. D 12. B,C 13. B 14. D 15. B 16. C
17. B 18. A 19. oceanic, neritic
20. temperate deciduous forest, taiga 21. tropical rain forest
22. secondary 23. decrease, decrease
24. Chaparral grows at lower temperatures and lower levels of precipitation.
25. a. Two species cannot occupy the same niche for long periods of time
    b. The data showed that the five different species of warblers were showing very little overlap in their feeding.

CHAPTER 26

1. between 1930-1935 2. until 1935 3. 1935 4. 1940
5. yes 6. a low number of animals that could reproduce
7. The animals could have damaged the environment and lowered the carrying capacity.
8. F 9. F 10. F 11. T
12. F 13. B 14. A 15. B 16. B 17. B
18. A 19. B 20. A 21. A 22. carrying capacity
23. slower, environmental resistance 24. genes
25. This is really what the entire chapter is about. Some reasons include the following. A female housefly couldn't acquire enough energy to lay all those eggs in a single year. If she laid them, abiotic and biotic factors would keep them all from hatching or maturing. Those that did survive would eventually die and decompose.

CHAPTER 27

1. Look at Figure 27.7 in the text. "A" should look like the one on the left (Sweden). "B" should look like the one on the left (Mexico). 2. a. 3 b. 7 c. 30 d. 18
3. F 4. F 5. T 6. T 7. F 8. C
9. C 10. A 11. A 12. B 13. A 14. C
15. A 16. A 17. E 18. bacteria, vaccine, smallpox
19. agricultural, increases 20. 40, 10.4-10.7 billion
21. Africa
22. a. *Any* rate of natural increase means that the population is growing, since it is saying that there are more births than deaths. Death rate must equal or exceed birth rate for the population to eventually stabilize or decrease.
    b. No

CHAPTER 28

1. the producers 2. 9.3 percent (from 450/4846)
3. The plants used some for their own needs. The herbivores could not utilize all of what they ate. Some plants died and were decomposed. It all went up as heat.

4. F 5. F 6. T 7. F 8. F 9. T
10. T 11. F 12. B 13. A 14. A 15. B
16. A 17. C 18. C 19. B 20. glycogen, fat
21. corn, wheat, potatoes, rice, 80,000
22. evaporation, pollution 23. solar, electricity
24. conservation, recycling, longer
25. There are two problems here. First, since it is public land, no one perceives themselves as owner, and is less likely to care for it. Second, each person stands to benefit if they cheat and add extra cattle to the land, however the cumulative effect is far too many cattle and overgrazing results.

CHAPTER 29

1. C,A,D,E,B,F 2. F 3. T 4. F 5. F
6. T 7. T 8. A,B,C 9. D 10. A,B 11. A,C
12. D 13. C 14. C 15. A 16. automobiles
17. ozone, ultraviolet 18. carbon monoxide, cigarette
19. nitric, sulfuric 20. less 21. sulfur oxides 22. threshold
23. The algae have low levels of DDT because it is a very long-lasting pesticide that is still active in their environment, and once they take it in during their normal metabolism it remains in their cells. The fish-eating birds have high levels because *they* ate many fish, which had already eaten many many fish which had eaten more algae than you could count. At each level, the DDT from everything eaten previously was held in the tissues of the topmost consumer, resulting in a concentrating effect.